The Evolution of Time: Studies of Time in Science, Anthropology, Theology

Edited by

Argyris Nicolaidis

Faculty of Science
Aristotle University of Thessaloniki
Thessaloniki
Greece

&

Wolfgang Achtner

Institute of Protestant Theology
Justus Liebig University
Giessen
Germany

CONTENTS

FOREWORD

What is it time? While all of us experience time, while everything we realize takes place within time, we are found in a difficult position, as we are reminded by Saint Augustine, to describe what time is. The eventual appreciations of time vary considerably. For some people time is an illusion, or simply a useful parameterization of the events. For other people time is the only reality, the generator and provider of everything.

Eleven scholars met twice in 2007, in order to address this thorny issue of time. A variety of opinions were presented, expressing the depth, the range and the intricacy of the time dynamics. The principal merit of these meetings consisted in bringing together colleagues from different disciplines. It involved scientists from the hard core of science (particle physics, relativity and cosmology), biologists and neurophysiologists, philosophers and theologians. The questions addressed include the notion of time in quantum mechanics and general relativity, the process of "self-organization" in time, the anthropic link of the external time to the human time, the biological time, the neurophysiology of time, the time during a mystical experience, the multiplicity of times and the universal description of time, our temporal existence and the eternal divinity, Kierkegaard's view on time and a comparison of related notions of time in philosophy and theology.

Is there a convergence among the different points of view? Is there a shared understanding of the notion of time? It is not that certain. Far from easily offered solutions, these proceedings respect the complexity of the issue, search for novel insights and bring forward the latest results from scientific research. For these reasons, the eBook is a trustworthy companion for an exciting trip in the land of time.

Christos Tsagas
Department of Astrophysics
University of Thessaloniki
Greece

PREFACE

Time, a fundamental component of human thought and experience, remains a most enigmatic and elusive one. A preeminent philosophical topic, time is linked to the dynamic interplay of being and becoming. Christian theology underlines the anthropological dimension of time, first analyzed by Saint Augustine. Natural Sciences study the different temporalities encountered in nature, from the vibrations of atoms to the planetary motions and the evolution of the universe itself. Neurosciences explore how parts of the human brain are associated to distinct ways of experiencing time.

The research workshop *"TIMES: Time in Science, Anthropology and Theology"* brought together scholars from the fields of Physics, Mathematics, Biology, Neuroscience, Psychology, Philosophy and Theology. The objective was to study the full dynamics of the time phenomenon and the evolution and higher level complexity of our own conceptions of time. A number of important questions were raised, with a broad range of answers indicated. Among them:

- The nature of time, *i.e.,* is time a fundamental notion shaping reality, or rather is it an illusion?

- What natural sciences tell us about time, especially theories like quantum mechanics, relativity and complex systems?

- The evolution of the universe involves different time scales. Billion of years for galaxy formation, million of years for the creation of biological entities, thousand of years for the human development. Is it possible to accommodate all these different time scales, within a single history in time?

- Is there any connection between the times and temporalities we encounter in nature, and the time conceived or sensed by the humans?

- Past – present – future, are these real divisions of time? Can we imagine a unification of the three domains?

- Apart from the standard causality, which leads us from the past to the future, can we consider other forms of causality in a time process? (*e.g.,* a dynamics where the present is conditioned by the future)

- Is eternity opposed to time, or may we consider the possibility of interference of time and eternity, the simultaneity of time and eternity?

- Can we encounter within a time process, the presence of human freedom, the existence of human being as a free agent?

These proceedings present each participant's contribution, drawn from their own field of knowledge and expertise. The individual contributions are grouped into broad chapters. The first chapter contains the contributions referring to, or inspired by physics. The second chapter includes the papers originating from biology and human sciences. The third final chapter contains the contributions from philosophy and theology.

In the first contribution, Prof. A. Nicolaidis considers the implications on our concept of time drawn from relativity theory, quantum mechanics and cosmology. The important element is evolution and time is the "all-begetting one", the essential condition for the realization of being. We are thus led to an "ontology in the temporal", inspired by the work of Peirce, Whitehead, Heidegger. A panorama of temporal structures, emerging in the different branches of science, is provided by Prof. K. Mainzer. The process of emergence of temporal patterns is contrasted to the philosophical tradition of "self-organization" (*autopoiesis*) as expressed by Aristotle, Kant, Shelling and to the theological tradition of revelation. Prof. P. Mittelstaedt focuses his attention on relativistic time, time as conceived within special relativity, general relativity, relativistic cosmology. The conditions for the existence of a universal cosmic time are studied and it is explored how the philosophical concept of eternity might emerge within the context of modern cosmology. Prof B. Carter invokes the anthropic reasoning to link the external time of the objective world to the characteristic time scale of human perception. He deduces that our capability for mental processing is favored when the ratio of gravitational to electric coupling is small.

Profs. G. Falkner and R. Falkner introduce us to the biological time, where time sustains qualitative alterations connected both to a memory of the past and an anticipation of the future. The adaptation process of an organismic self is exemplified with the case of the cyanobacteria during phosphate fluctuations. Prof. H. Förstl presents evidence that different parts of the brain are particularly relevant for certain aspects of time-related experience and behavior. Neurodegenerative diseases are approached as experiments of nature, indicating the extended brain areas, important for specific time-related tasks. Prof. U. Ott examines the time experience alterations during the mystical states of consciousness or meditation. The reported changes are characterized by a sense of timelessness and a feeling of all-encompassing unity. The monitoring of the mystic experience by EEG, measuring the electrical activity of the brain, reveals an increased activity in the gamma frequency range. Prof. J. Wackermann raises the important question how the multitude of "times" arising in diverse realms of reality, may lead to a universal description of time. It is proposed that inter-subjective synchronization and communication, allows the creation of a quasi-uniform, consistent time-keeping.

Prof. D. Evers brings in the philosophical and theological arguments regarding time and eternity. How to relate our temporal, transient existence to the eternal God? The ontological validity of time is rescued by creating forms and modes of human time, participants of God's eternal being. Prof. E. Gräb-Schmidt assumes the task of comparing the concept of time outlined by Malsburg, a physicist, to the model of time developed by the philosopher and theologian Kierkegaard. Malsburg argues for an eternal universe, bringing together past, present, future, while Kierkegaard upholds a creative present, which linked to the past, anticipates the future. The final contribution by Prof. W. Achtner offers a broad review of time in philosophy and theology (Plato, Plotinus, St. Augustine), in the mystical traditions of world religions (Islam, Hinduism, Buddhism), in neurophysiology, in the trinitarian concept of God.

Our first gathering in May 2007 was hosted by the Macedonian Museum of Contemporary Art (MMCA) in Thessaloniki, while our second gathering in September 2007 was hosted by the National Hellenic Research Foundation (NHRF) in Athens. Our meeting in MMCA ran in parallel with a major exhibition on "time and the times of the artwork", marking the 40th anniversary of the translation in Greek of Proust's novel *"À la recherche du temps perdu"*, with 27 artists presenting their own vision of time. The generated interaction between the participants of the workshop and the artists, their joint public debate on time, widened the horizon of cultural manifestations, revealing the different aspects of human intelligence, creativity, imagination. The cover of the proceedings portrays the work of one of the artists, Apostolos Georgiou. The title of the painting is "Five past three", and clearly indicates that time, rather than an objective or subjective abstract operation, is marked by concrete actions of human solidarity.

We would like to thank those who supported our endeavor: Prof. Xanthippe Heupel, president of MMCA, Prof. Denys Zacharopoulos, art director of MMCA, Prof. Dimitrios Kyriakidis, director and chairperson of NHRF, Dr. Karl-Heinz Thalmann, director of the Goethe Institute. The media sponsor of our public activities was the national TV channel ET3. All practical and organizational tasks were carried out thanks to the inventiveness and hard work of three young people: Dimitris Evangelinos, Marina Ntika, Panayiotis Tsalouhidis.

Last but not least, we would like to gratefully acknowledge the generous support of the John Templeton Foundation. The research workshop on Time was part of a major project, carried out by the European Research Network (ERN) "Science – Religion Interaction in the 21st Century", and coordinated by one of the editors (A.N.). A grant of the Templeton Foundation to ERN allowed us to organize research workshops, symposiums and public conferences. The present volume may be seen as a sign of our collective effort to respond to Sir John Templeton's invitation, to link the scientific achievements to the questions of value and meaning.

Argyris Nicolaidis
Faculty of Science
Aristotle University of Thessaloniki
Thessaloniki
Greece

Wolfgang Achtner
Institute of Protestant Theology
Justus Liebig University
Giessen
Germany

List of Contributors

Argyris Nicolaidis

Theoretical Physics Department, Aristotle University of Thessaloniki, Thessaloniki, Greece

Klaus Mainzer

Department of Philosophy and Theory of Science, Director of the Carl von Linde Academy, Technical University of Munich, Munich, Germany

Peter Mittelstaedt

Theoretical Physics Institute, University of Cologne, Cologne, Germany

Brandon Carter

Emeritus Director of Research, CNRS (National Centre for Scientific Research), Observatoire de Paris, Meudon, France

Gernot Falkner

Neurosignaling Unit, Cell Biology Department, University of Salzburg, Salzburg, Austria

Renate Falkner

Neurosignaling Unit, Cell Biology Department, University of Salzburg, Salzburg, Austria

Hans Förstl

Department of Psychiatry & Psychotherapy, Technical University Munich, Munich, Germany

Ulrich Ott

Bender Institute of Neuroimaging, Justus Liebig University Giessen, Giessen, Germany

Jiří Wackermann

Department of Empirical and Analytical Psychophysics, Institute for Frontier Areas of Psychology and Mental Health, Freiburg i. Br., Germany

Dirk Evers

Institute of Systematic Theology, Practical Theology and Religious Studies, Martin Luther Universitäat Halle-Wittenberg, Halle (Saale), Germany

Elisabeth Gräb-Schmidt

Institute of Ethics, Faculty of Protestant Theology, University of Tübingen, Tübingen, Germany

Wolfgang Achtner

Institute of Protestant Theology, Justus Liebig University, Giessen, Germany

2

Send Orders of Reprints at reprints@benthamscince.net

CHAPTER 1

The Dominion of Time

Argyris Nicolaidis[*]

Physics Department, Aristotle University of Thessaloniki, Greece

Abstract: We explore the meaning of time by studying the notions of time inherent in the different physical theories, the different representations we have built of Nature, notably: Classical Mechanics of Newton, Special and General Relativity of Einstein, Quantum Mechanics, Cosmology, Complex Systems, String Theory, Quantum Gravity. We notice that the notions of time, originating from the different physical theories, are not identical. They range from absolute, rigid specifications of time, to internal, relational considerations of time. Considering a physical system as an information-processing system, we link time to the generation of information and innovation. We contrast the concepts of time derived within Modern Science to Peirce's evolutionary metaphysics and Whitehead's process philosophy. We attempt to bypass the old dichotomy being-becoming, by indicating the need for an "ontology in the temporal", and suggesting temporality as a mode of existence (*τρόπος υπάρξεως*) of being.

Keywords: Process philosophy, relational time, modes of evolution, temporality, innovation time, Plato, Aristotle, Peirce, Whitehead, Newton, Einstein, quantum systems, cosmology, Christian theology.

1. INTRODUCTION

Time is a basic aspect of human experience, and yet as a notion, it is the most enigmatic and least tangible. Usually time is an implicit assumption in our discussion, a sort of a thread which links together the separate facts, the separate events into a narrative, a history, a continuum.

Time is closely linked to two fundamental notions that shape human thought, *being* and *becoming*. It is warranted to recall the insights of ancient Greeks who cherished such fundamental ontological issues.

For Plato, time is the mobile image of a motionless eternity (*κινητή εικόνα*

*Address correspondence to Argyris Nicolaidis: Theoretical Physics Department, Aristotle University of Thessaloniki, Greece; Tel: +30 2310 998143; Fax: +30 2310 998143; E-mail: nicolaid@auth.gr

Argyris Nicolaidis and Wolfgang Achtner (Eds)

αιώνος). It is a definition containing ambiguous terms and receiving multiple interpretations. I would like to stress in this view that time being an image, an icon (*εικών*), is linked to *εικώς λόγος*, that is logos based on experiences and conjectures (*εικασίες*) and not to the authentic transcendent logos. Thus for Plato time lacks an ontological foundation.

For Aristotle time is "the number of motion in respect of before and after" (*αριθμός κινήσεως κατά το πρότερον και ύστερον*). It is a definition which remains meaningful and valid for contemporary science. Essential characteristics of this operational definition are as follows:

- Time results from a mapping of motion into numbers. For each single event we associate a number. In mathematical notation, t [event] $= t_E$.

- We can establish an ordering in time, *i.e.,* for two events A, B we can quantify whether,

- $t_A < t_B$, $t_A > t_B$, or $t_A = t_B$.

- Numbers can be mapped onto a continuous line, and this operational time acquires the topology of a line. This Aristotelian approach brings time closer to space.

The most essential and yet practical question is how this mapping is achieved. Aristotle suggests the existence of "first motion", which serves as a measure for other types of motion. With the distinction between the imperfect sub-lunar world and the perfect supra-lunar world, first motion is attributed to celestial bodies of the supra-lunar world, which follow a perfect circular motion. Modern science questions such perfection in the heavens, thus, we may consider as first motion the oscillatory behavior of atomic particles, the atomic clocks.

Modern Physics was born in the 17th century, based on Cartesian premises and in total rejection of Aristotelian physics. Newton comments in his *Principia*: "I observe that ordinary people consider these quantities (time, space, motion) through their relationship to sensible objects. There are prejudices, and we have to distinguish the absolute from the relative, the true from the phenomenal, the mathematical from the common". Furthermore, Newton defines the absolute time: "Absolute, true and mathematical time, which flows equably without relation to anything external" [1]. Or, we could say that absolute time is what does change, when no change is noticed. Clearly Newton is in opposition to Aristotle who preceded him, and Einstein who followed him.

Einstein, in his Special Relativity, unified space and time into a continuum, space-time. He achieved this by raising the speed of light c, the speed of information transfer, to an absolute, a universal constant [2]. The different experiences in space-time are linked together and reflect a unique, concrete and unified nature. Following Einstein, the space-time is structured by the events taking place. The same event is described in different ways by different observers, thus obtaining a multiplicity of "personal languages". Thanks to the Lorentz transformations, there is a "dictionary", allowing "translation" from one language to another. Events connected together by causality belong to the same light cone, a cone in space-time bordered by signals travelling at the speed of light. Theoretical physicists consider two limiting cases regarding the value c of the speed of light:

i)　Everyday life, where all speeds are very small ($v \ll c$) and gravity is weak, is reproduced by the limit $c \rightarrow \infty$. In this limit the light cone expands, strict causality determines the whole space-time, and the Poincaré group is reduced to the Galilei group.

ii)　For very high energies and strong gravity, it has been shown [3], within string theory, that the appropriate physics is reproduced in the limit $c \rightarrow 0$. In this limit, the light cone shrinks, causality disappears and all events are un-correlated and spontaneous. The Poincaré group is reduced then to the Carroll group [4]. The group structure is named Carroll, to honor Lewis Carroll, the author of "Alice's Adventures in the Wonderland". And indeed in that limit, high energy - strong gravity, events would appear as spontaneous and uncorrelated, as in a Wonderland.

Einstein's General Theory of Relativity provides more radical changes. The space-time continuum acquires a dynamical character and its geometry is determined by the distribution of matter. The curvature of spacetime implies that the rate of change of time is not the same. A clock runs faster at the top of a skyscraper compared to a clock at ground level. The relativity aspect can be even more accentuated when we describe the same phenomenon by different accounts or histories. For example, time becomes frozen at the horizon of a black hole. While Newtonian time is external, spatially global, and unique, time in General Relativity is determined by the dynamics of the system itself and is neither spatially global nor unique. Kantian considerations of space and time as terms of an apriori intuition are undermined because space-time is represented by a particle-field, the graviton, much like all the other particle-fields. The whole space-time is under a continuous evolution, and General Relativity provides a framework for studies of the history of the universe.

A strong rift with the "Cartesian programme" (a definition of Husserl and Heidegger) occurs with the birth and development of quantum mechanics. The state of a quantum particle is represented by a linear superposition of two opposite eventualities A and $\bar{A}$ and a third term T serves as a bridge between them. The time evolution of the wave function of a particle takes place within an abstract Hilbert space. It is expected that after a period T the quantum system returns to its initial state. Thus the notion of a dynamical change is less pronounced in quantum theory. The outmost relational character of Quantum Mechanics has been recently fully established by showing that the relational logic of C.S. Peirce may serve as the conceptual foundation of the quantum theory [5].

We may view a physical system as an information-processing system, a system encoding and transferring information. The dynamical evolution of the system is represented by trajectories in the phase space Ω. At any given time t the actual partition of the phase space Ω by the system leads to the Shannon information $I(t)$. As time goes by we obtain different partitions of Ω, with increased (or decreased) occupation of the phase space, with finer (or coarser) distributions, thus leading to different information contents I. The overall information flow is determined by some global invariants, the so-called Lyapunov exponents λ_i.

$$I(t) - I(t = 0) = K \cdot t$$

where K is the sum of the Lyapunov exponents $\sum_i \lambda_i$.

The following cases are possible:

- $K = 0$, with $I(t) = I(t = 0)$, characterizes conservative systems with the phase volume and the information content of the system remaining constant. Example: The motion of the moon around the Earth. We can predict the place and the phase of moon for any time instant, and nothing new is expected.

- $K \neq 0$, with $I(t) \neq I(t = 0)$, refers to systems where the information content changes. Examples: A system coupled to a random external source may have $K > 0$ and the phase volume spreads over the entire phase space as t goes to infinity. A dissipative system, qualified by $K < 0$, has a shrinking phase volume and is asymptotically stable.

In the first case, time appears as the generator of the same patterns, and we may refer to a repetition time t_R. In the second case, time is linked to generation (or loss) of information, the emergence of new patterns, to creativity and innovation, and we may refer to an innovation time t_I, which is defined *via*:

$$t_I = \frac{I(t) - I(t = 0)}{K}$$

We propose then, that rather than mapping the motion into a line, we should map the motion into the **time plane**, the time plane formed by t_R and t_I.

We may consider the following cases (see Fig. **1**):

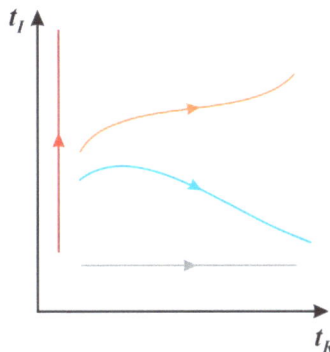

Figure 1: Evolutions in the time plane.

- Pure repetition: it is a trajectory parallel to the t_R axis.

- Pure innovation: it is a trajectory parallel to the t_I axis.

- A general dynamical process involving repetition and innovation is represented by a curve in the time plane, directed always to the right. Creation of information corresponds to increasing t_I, while information loss corresponds to decreasing t_I.

The evolution of the universe is a specific example with a tremendous increase of the information content. It is considered that the universe started with few bits at the Plank time, and presently it is estimated to contain 10^{120} bits. The universe is the ideal lab where we may study matter and the behavior of matter in diverse conditions, test the "unified models", probe the ultimate "laws" of Nature and even explore the nature of space-time. The current understanding of the evolution of the universe is impressive [6]. All major "cosmic events", dating from a few minutes after the "birth" of the Universe are well understood in the standard model of cosmology [7]. Within the inflationary scenario, the rich structure of the universe (galaxies, clusters of galaxies) is considered as a manifestation of quantum fluctuations in the inflaton field. Considering the random character of quantum fluctuations (analyzed by cosmic microwave background data), we understand that the current structure of the universe is the end result of *random* dynamics. It is noteworthy however, that the present universe also appears strangely unique, with a multitude of parameters having very specific values. This issue is analyzed usually within the "anthropic principle" framework [8, 9]. Should we then consider the possibility of "guided randomness"?

Trying to trace back the history of the universe to its very first instants, we deploy a theoretical arsenal enriched with models such as string theory, matrix models, extra dimensions, branes, non-commutative geometry, *etc*. In these models, the postulates of our current theories, like space-time, the number of dimensions and values of the coupling constants, are under question. Rather than being fixed, eternal quantities (or configurations), they appear to emerge as the result of a dynamics, where randomness is important. Thus time appears as an emerging entity, rather than as a permanent, fixed precondition.

Our journey through the different physical theories, trying to seize the nature of time, is revealing an important element: *evolution*. This manifest evolution transforms the universe from a homogeneous, amorphous object, to an heterogeneous, concrete, and unique entity, a *hypostasis*. We realize also that we have turned away from traditional ontological preoccupations. The classical Greek mind searched behind the appearance, the multiplicity, the change, behind the γίγνεσθαι–becoming, and looked for an invariant element, the first principle, the foundational είναι–being. In our case we are faced with the omni-presence of Time, time as "the all-begetting one", and *Becoming-ness* as the essential condition for the fulfilment of existing entities. These notions, the pre-eminence of Temporality and Becoming, may be linked to different traditions.

The Ionian philosopher Heraclitus (544-484 BC) should be credited for presenting the dynamics of opposites and continuous change as the permanent feature of the cosmos. In the biblical tradition we encounter also the importance of time. Truth is realized in time, as the continuous presence of God in the history of the world. But there are two great thinkers who provided an ontological foundation to time, Charles Sanders Peirce and Alfred North Whitehead.

Peirce ([10], p. 352), an ardent advocate of evolution, defined the three different modes of evolution:

1) *anancastic* evolution in which mechanical necessity reigns, the best example being Newtonian mechanics;

2) *tychastic* evolution, where chance predominates, Darwin's theory serving as an example;

3) *agapastic* evolution, where agape (αγάπη) is the source of creative growth and intelligible novelty.

Peirce was able to present a cosmological model in 1888, foreshadowing the developments in modern cosmology.

Our conceptions of the first stages of the development, before time yet existed, must be as vague and figurative as the expressions of the first chapter of Genesis. Out of the womb of indeterminacy we must say that there would have come something by the principle of *firstness*, which we may call a flash ([10], p. 245).

For Peirce, the universe is born out of *nothingness*, a "completely undetermined and dimensionless potentiality", which is characterized by freedom, chance and spontaneity. The initial flash was followed by other flashes until the events formed a quasi-continuous flow. Different streams of flashes might coalesce, while others are separated leading to many different worlds, which would know nothing of one another. *Secondness* is reached when a flash is genuinely second to another flash. Time is now supposed to emerge, as well the first germ of spatial extension. Habits will be formed in connecting certain states to certain others, and these habits will constitute a spatial continuum, but differing from our space by being very irregular in its connection. A bundle of formed habits will be a definite substance. For Peirce a substance exists as a precise web of relations. The permanence of some habits or tendencies will lead, *via* thirdness, to laws of nature.

For Whitehead [11] nature is a structure of evolving processes. The reality is the process. Whitehead proposed an ontology linked directly with time [12, 13]. He noticed that a steadily sounding note is explained as the outcome of vibrations in the air. He continued considering that the primordial elements of matter are the vibratory ebb and flow of an underlying energy, or activity. Accordingly there will be a definite period associated with each element. To the extent that a note of music is nothing at an instant, but requires a whole period in which to manifest itself, equally well the primordial element is nothing at any instant and needs its whole period in which to manifest itself.

These ideas of Whitehead strongly resonate with the most modern theory, string theory. In string theory, the "elementary" particles are not considered as autonomous, ever-lasting entities, but as special modes of excitation, or special modes of deformation of a string. Each string excitation has its own period, and the analogy with primordial elements of Whitehead is evident.

We are thus led to an "ontology in the temporal" [14] where time is the essential condition for the realization of being. Rather than referring to time, we should address and study different temporalities, where temporality is a specific way of organizing and deploying relations, structure, and patterns in time. We notice a transfer of ontological interest from "*τί εστί*" (what is it) to "*πώς εστίν*" (how is it). In other words, the category of relation or functionality takes precedence over that of essence (*ουσία*). Relational ontology is similarly supported by Christian theologians. A prime example is Saint Maximus the Confessor, who emphasized modes of existence (*τρόπος υπάρξεως*), underscoring the importance of relationship for both gnoseological and ontological purposes.

There are important issues not addressed in this paper such as the following:

i)   An approach to time is incomplete without examining the human experience of time. This is an old issue, first raised by St. Augustine, that is currently studied by neuroscience. It appears that parts of the human brain, the amygdala and neocortex, are the specific centers linked to the experience of time. This suggests the question "Can we connect temporalities observed in nature to the subjective human experience of time?

ii) In physics we are used to a causal principle where the future is conditioned by the past. In actual life we know that expectations, or anticipations of events to come in the future, structure our activities in the present. Can we include then in our considerations the notion of intention, finality, or *telos*? How firm is the distinction of past-present-future?

iii) Is there any way, while living in space and time, to go beyond the confines of time? Can we perceive time in a "timeless way" as suggested in religious experiences?

Our navigation into the ocean of time gives us a better understanding of the Aristotelian term "ka-i-ros" (*καιρός*). For Aristotle, *καιρός* is the appropiate-time, the suitable-time, when the best conditions for something unique to happen are met. I think then it is *καιρός* to assume the full potentialities of time.

ACKNOWLEDGEMENTS

Declared none.

CONFLICT OF INTEREST

The author(s) confirm that this chapter content has no conflict of interest.

DISCLOSURE

Part of information included in this chapter has been previously published in World Futures: The Journal of Global Education, Volume 64, Issue 2 (2008).

REFERENCES

[1]   I. Newton (1934) *Philosophiae Naturalis Principia Mathematica*, Book I, University of California Press, Berkeley.
[2]   A. Nicolaidis (2000) "Space, Time and Matter in Modern Physics," *Outopia*, p.77, May-June 2000 (in Greek).
[3]   H. De Vega and A. Nicolaidis (1992) "Strings in strong gravitational fields," *Phys. Lett.* 295B, 214.
[4]   J-M Levy-Leblond (1965) "Une nouvelle limite non-relativiste du groupe de Poincaré," *Ann. Inst. H. Poincaré*, 3A, 1.
[5]   A. Nicolaidis (2009) "Categorical foundation of Quantum Mechanics and String Theory", *Int. J. Mod. Phys.* A24, 1175-1183.
[6]   See for example A. Linde, "Particle Physics and Inflationary Cosmology", hep-th/0503203.
[7]   P. Peebles (1993) *Principles of Physical Cosmology*, Princeton University Press.
[8]   B. Carter (1974) "Large number coincidences and the Anthropic Principle in Cosmology," in *Confrontations of Cosmological Theories with Observational Data* (I.A.U. Symposium 63), ed. M. Longair, 291-298 (Reidel, Dordrecht, 1974); "The anthropic principle and its implications for biological evolution," *Philosophical Transactions, Royal Society of London*, A310: 347-363, 1983.
[9]   J. Barrow and F. Tipler (1986) *The Anthropic Cosmological Principle*, Oxford University Press.
[10] *The Essential Peirce* (1992) Vol. 1, ed. N. Houser and C. Kloesel, Indiana University Press.
[11] A. Whitehead (1997) *Science and the Modern World*, The Free Press.
[12] *Physics and the Ultimate Significance of Time* (1986) ed. D. R. Griffin, Albany: State University of New York Press.

[13]   *Physics and Whitehead: Quantum, Process and Experience* (2004) ed. T. E. Eastman and H. Keeton, Albany: State University of New York Press.

[14]   M. Heidegger (1962) *Being and Time*, Harper & Row; E. Levinas (1969) *Totality and Infinity: an Essay on Exteriority*, Duquesne University Press.

Send Orders of Reprints at reprints@benthamscince.net

CHAPTER 2

The Emergence of Temporal Structures in Complex Dynamical Systems

Klaus Mainzer[*]

Department of Philosophy and Theory of Science, Director of the Carl von Linde Academy, Technical University of Munich, Munich, Germany

Abstract: Dynamical systems in classical, relativistic and quantum physics are ruled by laws with time reversibility. *Complex dynamical systems* with time-irreversibility are known from thermodynamics, biological evolution, growth of organisms, brain research, aging of people, and historical processes in social sciences. Complex systems are systems that compromise many interacting parts with the ability to generate a new quality of macroscopic collective behavior the manifestations of which are the *spontaneous emergence of distinctive temporal, spatial or functional structures*. But, emergence means no mystery. Mathematically, the emergence of macroscopic features results from the nonlinear interactions of the elements in a complex system. Complex systems can also be simulated by computational systems. Thus, arrows of time and aging processes are not only subjective experiences or even contradictions to natural laws, but they can be explained by the nonlinear dynamics of complex systems. Human experiences and religious concepts of an arrow of time are considered in a modern scientific framework. Platonic ideas of eternity are at least understandable with respect to mathematical invariance and symmetry of physical laws. But Heraclit's world of change and dynamics can be mapped onto our daily real-life experiences of arrow of time.

Keywords: Symmetry of time, proper time, relativistic space-time, singularity, PCT-theorem, quantum cosmology, cosmic arrow of time, irreversibility in thermodynamics, evolutionary time, aging process, computational time, cellular automata, autopoiesis, self-organization, nonlinearity, eternity.

1. THE EMERGENCE OF TEMPORAL STRUCTURES IN CLASSICAL AND RELATIVISTIC DYNAMICS

According to Newton's laws of mechanics, a *dynamical system* is determined by a time-depending equation of motion. Newton distinguished *relative* and *absolute*

*****Address correspondence to Klaus Mainzer:** Department of Philosophy and Theory of Science, Director of the Carl von Linde Academy, Technical University of Munich, Arcisstrasse 21, D-80333 Munich, Germany; Tel: +49 89 289 25360; Fax: +49 89 289 25362; E-mail: mainzer@cvl-a.tum.de

time, assuming that all clocks of relative reference systems in the Universe could be synchronized to an absolute world-time of an absolute space. The *symmetry of time* is expressed by changing the sign of the *direction of motion* in an equation of motion. In *classical mechanics*, mechanical laws are preserved (invariant) with respect to all inertial systems moving uniformly relative to one another (Galilean invariance). A consequence of time symmetry is the conservation of energy in a dynamical system [1]. Newton's absolute space can actually be replaced by the class of inertial systems with Galilean invariance. But, according to the Galilean transformation of time, there is still Newton's distinguished absolute time in classical mechanics.

In 1905, Einstein assumed the *principle of special relativity* for all inertial systems satisfying the constancy c of the speed of the light ('Lorentz systems') and derived a common space-time of mechanics, electrodynamics, and optics. Their laws are invariant with respect to the Lorentz transformations. Time measurement becomes path-dependent, contrary to Newton's assumption of absolute time. Every inertial system has its *relative ('proper') time*. The situation is illustrated by the Twin paradox [2]. In a space-time system, twin brother A remains unaccelerated on his home planet, while twin brother B travels to a star with great speed. The traveling brother is still young upon his return, while the stay-at-home brother has become an old man. But, according to the *symmetry of time*, the twin brothers may also become younger. Thus, relativistic physics cannot explain the *aging of an organism* with direction of time. According to Einstein (1915), gravitational fields of masses and energies cause the *curvature of space-time*. Clocks are effected by gravitational fields: The gravitational red shift of a light beam in a gravitational field depends on its distance to the gravitational source and can be considered as dilatation of time. The effect is confirmed by atomic clocks.

Relativistic cosmology assumes an expanding universe in cosmic time [3]. According to Hubble's law of expansion, no galaxy is distinguished. The *Cosmological Principle* demands that galaxies are distributed spatially homogeneous and isotropic ('maximally symmetric') at any time in the expanding universe. In geometry, homogeneous and isotropic spaces have constant (flat, negative or positive) curvature. In two dimensions, they correspond to a Euclidean plane with flat curvature and infinite content, a negatively curved saddle, or a positively curved surface of a sphere with finite content. With the assumption of the Cosmological Principle and Einstein's theory of gravitation, H.P. Robertson and H.G. Walker derived the three standard models of an expanding universe with open cosmic time in the case of a flat or negative curvature and final collapse and end of time in the case of positive curvature. F. Hoyle's *steady state universe* (1948) without global temporal development can be excluded by overwhelming empirical confirmations of an expanding universe. K. Gödel's *travelling in the past* on closed world-lines in an anisotropic ('rotating') universe (1949) is excluded by the high confirmation of isotropy in the microwave background radiation.

The beginning and end of time gets new impact by the theory of *Black Holes* and *cosmic singularities*. According to the theory of general relativity, a star of great mass will collapse after the consumption of its nuclear energy. During 1965-1970, R. Penrose and S.W. Hawking proved that the collapse of these stars is continued to a point of singularity with infinite density and gravity. Thus, the singularity of a Black Hole is an *absolute end of temporal development*. The Schwarzschild-radius determines the event horizon of a Black Hole. Because of the *symmetry of time*, there might be also, White Holes with expanding world lines and exploding matter and energy, starting in a point of singularity. This idea inspired Hawkins's theorem of cosmic origin (1970): Under the assumption of the theory of general relativity and the observable distribution of matter, the universe has an *initial temporal singularity* ('*Big Bang*'), even without the additional assumption of the Cosmological Principle. Time is initialized in that point [4].

From different philosophical points of view, theists or atheists have supported or criticized the idea of an initial point of time, because it seems to suggest a creation of the universe. The mathematical disadvantage is obvious: In singularities of zero extension and infinite densities and potentials, computations must fail. Thus, nothing can be said about the origin of time in relativistic cosmology.

2. THE EMERGENCE OF TEMPORAL STRUCTURES IN QUANTUM DYNAMICS

According to Bohr's correspondence principle, a dynamical system of quantum mechanics can be introduced by analogy to a dynamical system of classical

(Hamiltonian) mechanics. Classical vectors like position or momentum are replaced by operators satisfying a non-commutative (non-classical) relation depending on Planck's constant h. The dynamics of quantum states is completely determined by time-depending equations (*e.g.*, Schrödinger equation) with *reversibility of time*. The laws of classical physics are invariant with respect to the symmetry transformations of time reversal (T), parity inversion (P), and charge conjugation (C). According to the *PCT-theorem*, the laws of quantum mechanics are invariant with respect to the combination PCT [5]. Thus, in spite of P-violation by weak interaction, the PCT-theorem still holds in quantum field theories. But, it is an open question how the observed violations of PC-symmetry and T-symmetry (*e.g.*, decay of Kaons) can be explained.

An immediate consequence of the non-commutative relations in quantum mechanics is Heisenberg's principle of uncertainty which is satisfied by conjugated quantities such as time and energy: Pairs of virtual particles and antiparticles can spontaneously be generated during a tiny interval of time ('*Planck-time*'), interact and disappear, if the product of the temporal interval and the energy of particles is smaller than Planck's constant. Thus, the quantum vacuum as the lowest energetic state of a quantum system is only empty of real particles, but full of virtual particles ('quantum fluctuations') [6].

Further on, according to Heisenberg's uncertainty principle, there are no time-depending orbits (trajectories) of quantum systems, depending on precise values of momentum and position like in classical physics. In order to determine the temporal development of a quantum system, R. Feynman suggested using the sum ('integral') of all its infinitely possible paths as probability functions. In *quantum cosmology*, the whole universe is considered a quantum system. Thus, Feynman's method of path integral can be applied to the whole universe. In this case, the quantum state (wave function) of the universe is the sum (integral) of all its possible temporal developments (curved space-times). In 1983, J. Hartle and Hawking suggested a class of curved space-times without singularities, in order to avoid the failure of relativistic laws in singularities and to make the cosmic dynamics completely computational [7]. Therefore, the real numbers of the time parameter in the corresponding path integral are replaced by imaginary numbers. Thus, in a Lorentz metrics $x^2 + y^2 + z^2 - t^2$ of space-time, the three coordinates x,

y, and z of space can no longer be distinguished from the time parameter t by a minus sign, because $- t^2$ is replaced by $- (it)^2 = - - t^2 = + t^2$ with imaginary number $i = \sqrt{-1}$. In that sense, the *early universe existed in a "timeless" (imaginary) state without beginning* until the *Big Bang* started its expansion [8].

According to Hawking's hypothesis of an *early universe without beginning*, Feynman's path integral allows different models of temporal expansion which are more or less probable – collapsing universes, critical universes, universes with fast (inflationary) expansion. They are all initiated by random quantum fluctuations in a quantum vacuum as ground state of a quantum system. The real time of a universe starts with its expansion. St. Augustine's famous question: "What did God do before the creation of our universe?" is now answered in the framework of quantum physics. In the beginning of the universe, there was a random quantum fluctuation of the quantum vacuum. In quantum physics, random events must not be determined by hidden causes without violating fundamental laws (EPR-experiment). Thus, in the beginning, God played dice to initiate our universe. If you don't believe in God, then, according to quantum cosmology, you must state: In the beginning, there was (quantum) *randomness*.

Hawking uses the (weak) Anthropic Principle to distinguish a universe like ours, enabling the evolution of galaxies, planets, and life, with an early inflation and later retarded expansion of flat curvature. From his hypothesis, R. Laflamme and G. Lyons [9] derived the forecast of tiny fluctuations of the microwave background radiation which was confirmed by the measurements of COBE in 1992. Thus, Hawking's hypothesis of an early universe without temporal beginning has been confirmed (until now), but not explained by an unified theory of quantum and relativistic physics which we still miss.

The temporal development of the universe can be explained by dynamics of phase transitions from an initial quantum state of high density to hot phase states of inflationary expansion and the generation of elementary particles, continued by the retarded expansion of galactic structures. Thus, the *emergence of cosmic structures* is made possible by phase transitions of the universe. Cosmic time is characterized by the development from a nearly uniform quantum state to more complex states of differing cosmic structures. In this way, we get a *cosmic arrow of time from simplicity to complexity*, which is characterized by a bifurcation scheme of global cosmic dynamics: An initial unified force has been separated step by step into the partial physical forces we can observe today in the universe: gravitation, strong, weak and electromagnetic interactions with their varieties of elementary particles [10].

If in the early universe gravitation and quantum physical forces are assumed to be unified, then we need a *unified theory of relativity and quantum mechanics* with new objects as common building blocks of the familiar elementary particles. The string theory [11] assumes tiny loops of 1-dimensional strings (10^{-35} m) with minimal oscillations generating the elementary particles. In a superstring theory, the unified early state corresponds to a transformation group of *supersymmetry*, which leaves the laws of the unified force invariant. During the cosmic expansion the early symmetry is broken into partial symmetries corresponding to different classes of particles and their interactions. Only three spatial dimensions of the more dimensional superstring theory are 'unfolded' and observable. Today, there are five 10-dimensional string theories and an 11-dimensional theory of supergravitation with common features ('dualities') and identical forecasts of the universe. They are assumed to be unifiable in the so-called M-theory. In this case, the *cosmic arrow of time* could be completely explained by phase transitions from simplicity to complexity. But we still must miss such a final explanation.

3.   THE   EMERGENCE   OF   TEMPORAL   STRUCTURES   IN THERMODYNAMICS

In physics, a direction of time was at first assumed in thermodynamical systems. According to R. Clausius, the change of the entropy S of a physical system during the time dt consists of the change d_eS of the entropy in the environment and the change d_iS of the intrinsic entropy in the system itself, *i.e.*, $dS = d_eS + d_iS$. For isolated systems with $d_eS = 0$, the *2nd law of thermodynamics* requires $d_iS \geq 0$ with increasing entropy ($d_iS > 0$) for *irreversible thermal processes* and $d_iS = 0$ for *reversible processes* in the case of thermal equilibrium. According to L. Boltzmann, entropy S is a measure of the probable distribution of microstates of elements (*e.g.*, molecules of a gas) of a system, generating a macrostate (*e.g.*, temperature of a gas): $S = k_B \ln W$ with k_B Boltzmann's constant and W number of probable distributions of microstates, generating a macrostate. According to the 2nd law, entropy is a measure of increasing disorder during the temporal development of isolated systems. The reversible process is extremely improbable. For Hawking, the cosmic arrow of the expanding universe from simplicity to complexity, from an initial uniform order to galactic diversity, is the true reason of the 2nd law.

Nevertheless, as the 2nd law is statistical and restricted to isolated systems, it allows the *emergence of order from disorder in complex dynamical systems* which are in energetic or material interaction with their environment (*e.g.*, convection rolls of Bénard-experiment, oscillating patterns of the Belousov-Zhabotinsky-Reaction, weather and climate dynamics) [12]. In general, the development of dissipative systems can be characterized by *pattern formation of attractors* (*e.g.*, fixed point attractor, oscillation, chaos) and *temporal bifurcation trees*. In a critical distance to a point of equilibrium, the thermodynamical branch of minimal production of energy ('linear thermodynamics') becomes instable and bifurcates spontaneously into new locally stable states of order ('symmetry breaking') [13]. Then, the nonlinear thermodynamics of nonequilibrium starts. If the system is driven further and further away from thermal equilibrium, a bifurcation tree with nodes of locally stable states of order is generated. Global *pattern formation and emergence* of complex dynamical systems can be *irreversible*, although the laws of locally interacting elements (*e.g.*, collision laws of molecules in a fluid) are *time-reversible*.

4.   THE EMERGENCE OF TEMPORAL STRUCTURES IN EVOLUTIONARY DYNAMICS

The *emergence of life* is not so special in the universe. It is explained by the nonlinear dynamics of complex molecular systems. Catalytic hypercycles are prototypes of complex molecular systems with catalytic and autocatalytic feedback loops of nonlinearly interacting molecules. In a prebiotic evolution, self-assembling molecular systems become capable of self-replication, metabolism, and mutation in a given set of planetary conditions. It is still a challenge of biochemistry to find the molecular programs of generating life from 'dead' matter. *Darwin's evolution of species*, as far as it is known on Earth, can also be characterized by *phase transitions* and *temporal bifurcation trees*. Mutations are random fluctuations in the bifurcating nodes of the evolutionary tree, breaking the local stability of a species. Selections are the driving forces of branches, leading to further species with local stability. The distance of sequential species is determined by the number of genetic changes. *Evolutionary time* can be measured on different scaling, *e.g.*, by the distance of sequential species and the number of sequential generations of populations. Its *temporal direction* is given by the order of ancestors and descendants.

As conditions changed in the course of the Earth's history, *complex cellular organisms* have come into existence, while others have died out. Entire populations come to life, mature, and die, and in this they are like individual organisms. But while the sequence of generations surely represents the time arrow of life, many other distinct biological time rhythms are discernable. These rhythms are superimposed in *complex hierarchies of time scales*. Each hierarchical layer *emerges* from the previous ones by phase transions. They include the temporal rhythms of individual organisms, ranging from biochemical reaction times to heartbeats to jet lag, as well as the geological and cosmic rhythms of ecosystems.

Complex systems that consist of many interacting elements, such as gases and liquids, or organisms and populations, may exhibit separate temporal developments in each of their numerous component systems. The complete state of a complex system is therefore determined by statistical distribution functions of many individual states. It has been proposed by B. Misra, I. Prigogine *et al.* that time can be defined as an operator describing changes in the complete states of complex systems. This *time operator* would then represent the average *age* of the different system components, each in its distinct stage of development [14]. For example, a 50-year-old person could have the heart of a 40-year-old one, but, as a smoker, the lungs of a 90-year-old one. Organs, arteries, bones, and muscles are in distinct states, each according to its particular condition and genetic predisposition. The time operator is thus intended to indicate the *irreversible aging of a complex system*, its *inner* or *intrinsic time*, not the *external and reversible clock time*. In short: we distinguish the irreversible *operator time* of complex dissipative systems and the reversible *parameter time*. Operator time refers to the *spontaneous emergence of distinctive temporal, spatial or functional structures in complex systems* [15].

The *human brain* may also be regarded as a complex system in which many neurons and different regions of the brain interact chemically and constitute collective cell assemblies by synchronously firing states. The *emergence of mental states* is explained by the neural correlates of firing cell assemblies. Our individual experiences of "*duration*" and "*aging*" are related to the operator time of complex-system states in the brain, depending on different sensory stimuli, emotional states, memories, and physiological processes. Hence, our *subjective awareness* of time does not contradict to the laws of science, but is explained by the dynamics of a complex system. Our intimate subjectively experienced flow of time was described in many examples of literature and poetry. The neurobiological knowledge of brain dynamics does not turn us into a Shakespeare or a Mozart. In this sense, natural sciences and humanities are complementary.

In econophysics, the theory of complex systems is also applied to the *temporal dynamics of socio-economic systems*. A city, for example, is a complex residential region in which different districts and buildings have distinct traditions and histories. Again, its development is explained by the *emergence of distinctive temporal, spatial or functional structures in complex systems*. New York, Brasilia, and Rome are the result of distinct temporal development processes. The time operator of a city refers literally to the average age of many distinct stages and styles of development. Institutions, states, and cultures are characterized by growth and aging processes with typical operator time. Today, there is the dramatic problem of *aging societies* in highly developed countries (*e.g.*, Japan, Germany). From the point of view of complex dynamical systems, the discussions of age is not just metaphorical, but can be explained in terms of structural dynamics [16].

5. THE EMERGENCE OF TEMPORAL STRUCTURES IN COMPUTATIONAL AND INFORMATION DYNAMICS

Modern technical societies depend sensitively on the capacities of computers and information networks. *Computational time* is a measure of the time needed to solve a problem by a computer. As a measure of a problem's complexity, one focuses on the running time and data storage requirements of an algorithm and their dependence on the length of the input. The theory of computational complexity deals with the classification of problems into complexity classes depending on running time and input length. It is suspected that extremely shorter computational times are possible with quantum computers operating with quantum dynamics. Nevertheless, classical computers as well as quantum computers are based on the concept of *time reversibility*: The laws of nature under which they operate permit, in principle, their computing processes (other than the act of measurement and reading out in the case of quantum computers) to run backward in time.

The question arises whether it might also be possible to use computers simulating *time-irreversible processes* that are well known from biological evolution and the self-organization of the brain [17]. The emergence of cellular patterns was simulated for the first time in the 1950s by von Neumann's *cellular automata*. Computer experiments show the *emergence of patterns* that are familiar as the attractors of complex dynamic systems [18]. There are oscillating patterns of reversible automata and irreversible developments from initial states to final patterns. For example, in the case of a fixed point attractor, all developments of a cellular automaton develop to the equilibrium state of a fixed pattern which does not change in the future. As these developments are independent of their initial states, they cannot be reconstructed from the final equilibrium state.

Further on, there are cellular automata *without long-term predictions* of their time-depending pattern formation. These are cellular automata with the property of universal computability. *Universal computation* is a remarkable concept of *computational complexity* which dates back to Alan Turing's universal machine. A universal Turing machine can by definition simulate any Turing machine. According to the Church-Turing thesis, any algorithm or effective procedure can be realized by a Turing machine. Now Turing's famous Halting problem comes in. Following his proof, there is no algorithm which can decide for an arbitrary computer program and initial condition if it will stop or not in the long run. Consequently, for a system with universal computation (in the sense of a universal Turing machine), we cannot predict if it will stop in the long run or not. Assume that we were able to do that. Then, in the case of a universal Turing machine, we could also decide whether any Turing machine (which can be simulated by the universal machine) would stop or not. That is obviously a contradiction to Turing's result of the Halting problem. Thus, systems with universal computation are unpredictable. Unpredictability is obviously a high degree of complexity. It is absolutely amazing that systems with simple rules of behavior like cellular automata which can be understood by any child lead to complex dynamics which is no longer predictable.

There are at least some few cellular automata which definitely are Universal Turing machines. It demonstrates a striking *analogy of natural and computational processes* that even with simple initial conditions and locally reversible rules many dynamical systems can produce globally complex processes which cannot be predicted in the long run.

The paradigms of parallelism and connectivity are of current interest to engineers engaged in the design of *neurocomputers* and *neural networks*. They also work with simple rules of neural weighting simulating local connectivity of neurons in living brains. Patterns of neural self-assemblies are correlated with cognitive states. With simple local rules neural networks can produce complex behavior, again. In principle, it cannot be excluded that this approach will result in a technically feasible neural self-organization leading to systems with consciousness, and specifically with time awareness. Thus, computational systems (*e.g.*, robots) with spontaneously emerging features are possible.

In a technical co-evolution, *global communication networks* of mankind have *emerged* with similarity to self-organizing neural networks of the brain. Data traffic of the Internet is constructed by data packets with source and destination addresses. Local nodes of the net ('routers') determine the local path of each packet by using weighting tables with cost metrics for neighboring routers. There is no central supervisor, but there are only local rules of connectivity which can be compared to self-assembling neural nets. Buffering, sending, and resending activities of routers can cause high densities of data traffic spreading over the net with patterns of oscillation, congestion, and even chaos. Thus, again, simple local rules generate the *emergence of complex patterns of global behavior*.

Global information networks store millions of human information traces. They are *information memories* of human history, reflecting the *aging process of mankind as a complex dynamical system*. What is the future of mankind and its information systems in the universe? Cosmic evolution can also be considered as the aging process of a complex dynamical system. If we are living in a flat universe according to recent measurements, then relativistic cosmology forecasts an infinite expansion into the void with increasing dilution of energy and decay in Black Holes. Does it mean the decay of all information storages and memories of the past, including mankind, an *aging universe* with 'Cosmic Alzheimer'? Or may we believe in the fractal system of a bifurcating multiverse with the birth and recreation of new expanding universes? As far as we know there is a *cosmic arrow of time* in our universe, but it is still open where it is pointing at. With that question, we pass the boundary from science to religion. From a philosophical point of view, science is still following the traces of pre-Socratic philosophers starting in Greece in the 5th century B.C.: Is the world in an irreversible change without beginning and ending in the sense of Heraclitus or is, according to Parmenides, only being real and change an illusion.

6. THE EMERGENCE OF TEMPORAL STRUCTURES FROM A PHILOSOPHICAL AND THEOLOGICAL POINT OF VIEW

The concept of spontaneous emergence of distinctive temporal, spatial or functional structures has an old philosophical tradition. In Aristotelian physics emergence means a teleological process transforming structureless matter into form. The growth of a plant or an organism from a seed or egg to its mature or adult form is a typical Aristotelian example of emergence. In the scholastic philosophy of the middle ages, the transformation from structureless matter to complete forms was called an actualization of potentiality. In modern times Kant criticized the ontological and teleological approach of Aristotelian tradition, because, following Newtonian mechanics, nature can only be explained by causal forces and not by goals and functions. Nevertheless, Kant recognized that classical mechanics failed in explaining the emergence of life. Thus, in his *Critique of Judgment*, he accepted teleological models as metaphoric interpretations of life, but not as physical explanations. For the first time, emergence of structure was described by a kind of "*self-organization*" (*autopoiesis*). But Kant proclaimed in a famous quotation: "The Newton of a blade of grass must still be found".

In the early 19th century, the ideas of self-organization (*autopoiesis*) were revisited during the German period of romantic literature and romantic natural philosophy. Inspired by the new ideas of electromagnetism, Schelling (1775-1854) explained emergence and self-organization by cyclic causality of opposite forces. In the tradition of scholasticism, he distinguished producing and emerging processes in nature (*natura naturans*) and the generated products of objects and organisms (*natura naturata*). *Emergence* and *self-organization* refer to *natura naturans*, but the *emerged products* of nature to *natura naturata*. Although Schelling's natural philosophy influenced famous physicists of his days (*e.g.*, H.C. Oerstedt, J.W. Ritter), he was also attacked by mathematicians and mathematical physicists, because "*emergence*" and "*self-organization*" seemed to be only idealistic speculations which could be neither experimentally analyzed nor mathematically formalized in those days.

In modern systems theory, the scientific situation has completely changed. *Emergence* and *self-organization* can be mathematized by nonlinear dynamics of complex systems [19]. Schelling's cyclic causality is made precise by mathematical nonlinearity. There are stochastic differential equations (*e.g.*, master equation) modeling phase transitions of complex systems. Phase transitions refer to steps of emergence. *Complexity* has become an interdisciplinary (or "transdisciplinary") topic cutting across all traditional disciplines of the natural and life sciences, engineering, economics, medicine, neuroscience, social and computer science. Kant's "*Newton of a blade of grass*" is found: Models of such systems can be successfully mapped onto quite diverse real-life situations like the climate, the coherent emission of light from lasers, chemical reaction-diffusion systems, biological cellular networks, the dynamics of stock markets and of the internet, earthquake statistics and prediction, freeway traffic, the human brain, or the formation of opinions in social systems, to name just some of successful applications. Although their scope and methodologies overlap somewhat, one can distinguish the following main concepts and tools of nonlinear systems theory: self-organization, emergence, nonlinear dynamics, synergetics, turbulence, dynamical systems, catastrophes, instabilities, stochastic processes, chaos, graphs and networks, cellular automata, adaptive systems, genetic algorithms and computational intelligence.

The *emergence of temporal structures* is explained by phase transitions of complex systems. Structures, organisms, and organizations have their own proper time and scaling. Thus, every individual being has its proper life time. Their time-depending trajectories are connected in a complex network of evolution and human history. Time is not only external clock-time, but a complex network of internal systems dynamics cutting across all layers of being (operator time). The individual time of a human being is embedded in the history of mankind which is part of cosmic time. In *theology*, birth, growth and death are the fundamental experiences of extential human time which are distinguished from God's eternity. *Eternity* is often defined as absence of time or a timeless state. But, according to Hawking's model of an initial universe without beginning, a timeless state can be mathematized by a path integral.

Timelessness is not sufficient to characterize eternity. From a theological point of view, eternity does not mean a physical state, but is related to the sense of human life. Is our individual life only a statistical trajectory in a probabilistic distribution of mankind evolving according to the stochastic laws of nonlinear dynamics? In the Jewish-Christian tradition, each unknown man in past and future is not forgotten or unimportant but personally loved and accepted. Eternity means that each human life is embedded in a *sense beyond the finite time of an individual life*. In the old Latin translation of Psalm 30,16 the *eternal sense of life* is described by a simple, but wonderful picture: "*In manibus tuis tempora mea*" (My times are in your hands). These simple words express more than all abstract formulations of theological and philosophical eschatologies. Obviously, they cannot be derived from mathematical theories, but need another source of *emergence* which we call *revelation*.

ACKNOWLEDGEMENTS

Declared none.

CONFLICT OF INTEREST

The author(s) confirm that this chapter content has no conflict of interest.

DISCLOSURE

Part of information included in this chapter has been previously published in Foundation of Physics, Volume 40, Numbers 9-10 (2010), pp 1638-1650.

REFERENCES

[1]   Noether, E. (1918). "Invariante Variationsprobleme", Nachr. Ges. Wiss. Göttingen, Math.-Phys. Kl., pp. 235-257.
[2]   Illustrated in Audretsch, J. and Mainzer, K. (eds.) (1994), "Philosophie und Physik der Raum-Zeit", Mannheim, B.I. Wissenschaftsverlag 2nd Edition, pp. 41, 60.
[3]   Cf. Audretsch, J. and Mainzer, K. (eds.) (1990), "Vom Anfang der Welt. Wissenschaft, Philosophie, Religion, Mythos", München, C.H. Beck 2nd Edition.
[4]   Hawking, S.W. and Ellis, G.F.R. (1973), "The Large Scale Structure of Space-Time", Cambridge, Cambridge University Press; Hawking, S.W. (1988), "A Brief History of Time: From the Big Bang to Black Holes", London, Bantam Books.
[5]   Cf. Pauli W. (1957), "Niels Bohr and the Development of Physics", Pergamon Press, London.
[6]   Cf. Audretsch, J. and Mainzer, K. (1996), "Wieviele Leben hat Schrödingers Katze? Zur Physik und Philosophie der Quantenmechanik", Spektrum Akademischer Verlag, Heidelberg.
[7]   Hartle, S. and Hawking, S.W. (1983), "Wave Function of the Universe", Phys. Rev. D 28, pp. 2960–2975.
[8]   Mainzer, K. (2000), *Hawking*, Freiburg, Herder, p. 81.
[9]   Hawking S.W., Laflamme, R. and Lyons, G.W. "The Origin of Time Asymmetry", Phys. Rev. D47 1993, pp. 5342-5356.
[10]   Cf. Mainzer, K. (1996), "Symmetries of Nature", Berlin, New York, De Gruyter.
[11]   Greene, B. (1999), "The Elegant Universe: Superstrings, Hidden Dimensions, and the Quest for the Ultimate Theory", New York, W.W. Norton & Company.
[12]   Cf. Mainzer, K. (2007), "Thinking in Complexity. The Computational Dynamics of Matter, Mind, and Mankind", Berlin, Heidelberg, New York, Springer 5th edition.
[13]   Haken H. and Mikhailov, A. (eds.) (1993), "Interdisciplinary Approaches to Nonlinear Complex Systems", Berlin, Springer.
[14]   Misra, B. and Prigogine, I. (1982), "Geodesic instability and internal time in relativistic cosmology", Phys. Rev. D 25, pp. 921–929.
[15]   Prigogine, I. (1979), "From Being to Becoming – Time and Complexity in Physical Sciences", San Francisco, W.H. Freeman & Co; Mainzer, K. (2002), "The Little Book of Time", New York, Copernicus Books.
[16]   Mainzer, K. (2007) "Thinking in Complexity", ref. [12], Chapter 7.
[17]   Mainzer, K. (2007), "Thinking in Complexity", ref. [12], Chapter 6.
[18]   Wolfram, S. (2002), "A New Kind of Science", Champaign, Wolfram Media Inc.
[19]   Mainzer, K. (2005), "Symmetry and Complexity. The Spirit and Beauty of Nonlinear Science", Singapore, World Scientific.

Send Orders of Reprints at reprints@benthamscince.net

CHAPTER 3

Cosmic Time and the Evolution of the Universe

Peter Mittelstaedt[*]

Theoretical Physics Institute, University of Cologne, Cologne, Germany

Abstract: Within the framework of General Relativity, a universal cosmic time that is relevant for all observers who are commoving with the cosmic substratum can be established by a convenient system of coordinates. However, this is only possible, if the substratum is free from vortices, in accordance with the so-called "Cosmological Principle". Hence, the existence of a cosmic time depends on contingent properties of the cosmic substratum. – Only if these properties are given, the age and the dynamical behavior of the universe can consistently be described. Under this assumption, we discuss the question, whether the universe could have an infinite past and show that – in spite of a long lasting philosophical debate from Proclus to Kant – there are consistent physical models with infinite age. Furthermore, we argue that the concept of eternity, as it was conceived in the philosophical tradition, can be given an adequate meaning within the context of modern relativistic cosmology. Also this problem can be traced back to ancient Greek philosophy and will be discussed in detail.

Keywords: Cosmic substratum, incoherent matter, cosmological principle, cosmic time, finite past, big bang models, infinite past, Friedmann equation, Eddington-Lema^itre universe, Penrose diagrams, infinitely extended global systems.

1. INTRODUCTION

The concept of absolute time that was conceived by Newton in his "Principia" of 1687 [1], was one of the most important constituents of the conceptual framework of physics in the following 200 years. In the "Principia" Newton wrote [2]:

> *"Absolute, true, and mathematical time, of itself, and from its own nature flows equably without relation to anything external",*

or, in the original Latin formulation:

*Address correspondence to Peter Mittelstaedt:** Theoretical Physics Institute, University of Cologne, D-50937 Cologne, Germany; Tel: +49 (0) 221 470 3480, 4306; Fax: +49 (0) 221 470 5169; E-mail: mitt@thp.uni-koeln.de

"tempus absolutum, verum et mathematicum, in se et natura sua absque relatione ad externum quodvis, aequabiliter fluit".

According to Newton, absolute time is a real entity that we know, however, only partially. Newton did not claim, that absolute time is directly measurable and he even admitted that "it is possible that there is no uniform motion by which, time may have an exact measure" [3]. Nevertheless, he believed in the existence of absolute time at least for two reasons. First, he was convinced to have proved the existence of absolute motion by the rotating-bucket experiment, and as consequence the real existence of absolute time and absolute space. Second, his imperturbable belief in the existence of absolute time is religiously motivated as can be seen in the second and third edition of the Principia [4].

Two hundred years later, in the 19th century, the concept of absolute time was criticized very seriously by Ernst Mach and Henri Poincaré. Ernst Mach considered absolute time as a purely metaphysical concept which does not appear in any realizable situation. In his "Mechanik" [5], he could show that Newton's interpretation of the realizable rotating-bucket experiment is a fallacy and does not demonstrate the existence of absolute motion. Consequently, also absolute time does not exist as an empirical concept. In addition, Ernst Mach could show that there are no motions that can be used as reliable and universal clocks. Which kind of motion we consider to be a clock is merely a matter of convention. In summarizing these results, Mach argued that absolute time is a useless (müßiger) metaphysical concept and should be completely eliminated in physics.

On the basis of similar considerations, Henri Poincaré arrived at the result that the measure of time that we use in physics is not based on empirical grounds but on arbitrary conventions and on non-empirical principles like simplicity. Within the framework of classical mechanics, Poincaré could explicitly apply these ideas and requirements [6]. Another aspect of time, which was first clarified by Poincaré, is the concept of distant simultaneity. Using an astrophysical example, Poincaré could show that any concept of simultaneity of events separated in space is based on conventional stipulations that are – in most cases – tacitly presupposed in our physical theories.

In Newton's mechanics, all the problems mentioned disappear. Absolute time – if it existed – would determine the direction of time, the measure of time, motions with constant velocity, and an applicable concept of distant simultaneity. However, the rigorous elimination of absolute time in physics, as it was required by Mach and Poincaré, leads to a disastrous situation, since it would take away from physics one of the most important basic concept without offering an adequate substitute. A few years later, a very convenient and useful substitute for the old concept of absolute time was introduced by Einstein 1905 in his *Theory of Special Relativity* and 1916 in his *General Theory of Relativity.*

2. RELATIVISTIC TIME

2.1. Special Relativity

Without the hypothesis of an absolute and universal time, every observer who is equipped with rods and clocks has his own private time. As in Newton's space-time, we can start also here with the concept of an inertial system. A system of inertia is a frame of reference such that the trajectory of a force-free test body is given by a straight line in the three dimensional space. Different systems of inertia $I, I', I'', \ldots$ have different time coordinates $t, t', t'' \ldots$ that can no longer be identified as in the Newtonian theory of space-time. However, there are well known coordinate transformations – the Lorentz-transformations – that connect the coordinates of different systems of inertia I and I'.

An inertial observer is usually equipped with clocks and rods for measuring intervals of the coordinate time t and distances of the space coordinate x. If two observers $\mathcal{O}_1$ and $\mathcal{O}_2$ have different positions x_1 and x_2 in space and are, relatively to each other, at rest then the question arises, how the local clocks $\mathcal{C}_1$ and $\mathcal{C}_2$ of $\mathcal{O}_1$ and $\mathcal{O}_2$, respectively, can be synchronized. The method, first introduced by Einstein, defines distant simultaneity by means of light signals. This procedure is not uniquely defined since it contains still a conventional assumption about the propagation of light. The most simple convention is, in an inertial system to choose the space-time coordinates such, that the one-way velocity of light is isotropic. It leads to the well known Einstein-synchronization [7]. A light signal that is emitted by observer $\mathcal{O}_1$ with coordinates (x_1, t_1) arrives the other observer $\mathcal{O}_2$ at (x_2, t_2). If it is reflected immediately, it arrives at $\mathcal{O}_1$ at (x_1, t_3), say. The observer O_1 can read only his local clock C_1, *i.e.,* he knows only the values of emission and absorption t_1 and t_3, respectively. However, the assumption of isotropy implies that the one-way velocity of light is the same in both directions [8]. Hence, the time t_2 of reflection reads:

$$t_2 = t_1 + \tfrac{1}{2}(t_3 - t_1),$$

in accordance with the experimental set-up shown in Fig. **1**.

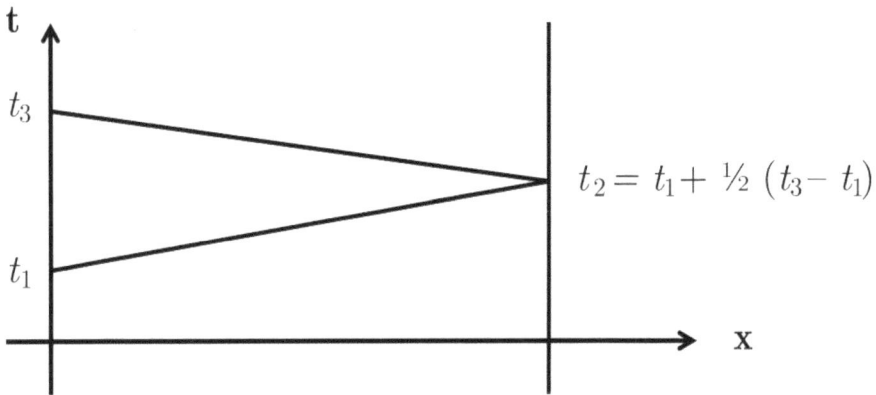

Figure 1: Einstein's definition of distant simultaneity.

Formally, the space-time manifold of Special Relativity is described by a Minkowskian space, *i.e.,* by a four dimensional pseudo-Euclidean space with signature 2. In an inertial system we can always introduce Cartesian coordinates x^k and a time coordinate $x^0 = ct$ such that force-free test bodies move not only on straight lines but also with constant velocity and that the propagation of light is isotropic. In these coordinates, the metric tensor $\eta_{\mu\nu}$ is diagonal and the line element reads [9]:

$$ds^2 = \eta_{\mu\nu} dx^\mu dx^\nu = \left(dx^0\right)^2 - \left(dx^1\right)^2 - \left(dx^2\right)^2 - \left(dx^3\right)^2.$$

Within the framework of the Minkowskian space-time there is no problem to describe also accelerated frames of reference. For an accelerated observer a local clock is defined by a so-called "standard clock", *i.e.,* a clock that cannot be affected by inertial forces. A standard clock measures a *proper time* interval $d\tau$ defined by $dt^2 = ds^2 / c^2$, instead of an interval dt of the *coordinate time* [10].

Special Relativity can be applied to all physical phenomena, provided that there are no gravitational fields. The reason is that in the presence of gravitation, systems of inertia do not exist. The incorporation of gravitational fields requires a complete reformulation of the theory of space-time. The resulting theory of space, time, and gravity is the General Theory of Relativity, which was formulated by Einstein in 1916.

2.2. General Relativity

If gravitation is taken into account, the space-time continuum is no longer described by a Minkowskian space but by a four–dimensional pseudo–Riemannian space with signature 2. In a Riemannian space–time force–free test particles don't move any longer along straight lines but on geodesic lines of the four–dimensional space–time. The matter will be described here by the energy-momentum tensor $T^{\mu\nu}$. The connection between the structure of the Riemannian space-time and the tensor $T^{\mu\nu}$ of the matter is determined by Einstein's field equations:

$$G^{\mu\nu} := R^{\mu\nu} - \tfrac{1}{2} R \, g^{\mu\nu} + \Lambda g^{\mu\nu} = -\kappa T^{\mu\nu}.$$

The left hand side of this equation contains geometrical quantities, the metric tensor $g^{\mu\nu}$, the Ricci-tensor $R^{\mu\nu}$ and its trace $R = R^{\mu\nu} g_{\mu\nu}$, whereas the right hand side consists of the matter tensor $T^{\mu\nu}$. (Λ is the cosmological constant, κ Einstein's gravitational constant $\kappa = 8\pi \, k/c^4$, and k Newton's gravitational constant). The tensor $G^{\mu\nu}$ is often called "Einstein tensor". Formally, it can be seen that the covariant derivative $G^{\mu\nu}{}_{;\nu}$ of the Einstein tensor vanishes identically and thus, according to Einstein's field equations, also the covariant derivative $T^{\mu\nu}_{;\nu} = 0$ of the matter tensor vanishes. Since the equation $T^{\mu\nu}_{;\nu} = 0$ agrees with the equation of motion mentioned above, we find that Einstein's field equations imply the equations of motion of the field creating matter.

Hence, the geometry of the Riemannian space-time determines the motion of matter and the behavior of clocks. Conversely, the distribution of matter, given by a tensor $T^{\mu\nu}$ determines the structure of space–time. Of course, it must be emphasized, that not the matter alone determines the geometry of the Riemannian space–time, but only the matter together with convenient boundary conditions for the gravitational field. In case of matter that is finitely extended in space, only very few *exact* solutions of Einstein's field equations are known, *e.g.,* the famous Schwarzschild solution. This solution was of particular importance for the acceptance of General Relativity in the scientific community, since for this model the first numerical results of the theory could be derived.

3. RELATIVISTIC COSMOLOGY

Relativistic cosmology is based on Einstein's field equations, which connect matter and energy with the metric of the four-dimensional space-time continuum. The metric is usually expressed by the line element [11]:

$$ds^2 = g_{\mu\nu}dx^\mu dx^\nu.$$

Except from a few exact solutions for local distributions of matter that are known today, Einstein's equations possess also global solutions, that are not restricted to a finite region of space-time. These global solutions that grasp the entire space-time are often considered as models of the universe. There are a huge number of exact global solutions of Einstein's field equations that differ by the matter model, the invariance properties, the boundary-and initial conditions, and by the large-scale topology. Famous global solutions of the field equations with the simplest model for the distribution of matter are: the constant Einstein solution of 1917, the expanding Friedmann solution of 1922, and the rotating Gödel solution of 1949.

Here, we are interested in global solutions of the field equations. For the matter we will use always the most simple model of incoherent matter without pressure, *i.e.,* dust. The geometry of the four dimensional space–time that is generated by this matter *via* its gravitational field, is the geometry of a four–dimensional pseudo-Riemannian space–time with signature two. This geometry is characterized by a space–time manifold $\mathcal{M}$ and the metric g, which can be determined in any special situation by means of Einstein's equations.

The elements of the cosmic substratum, the incoherent matter, are moving along trajectories that are time-like "geodesic lines" of the same Riemannian space-time that is generated by the cosmic substratum in question. Observers, who are located on elements of the cosmic substratum are locally "freely falling" observers who are not affected by inertial or gravitational forces. We will assume here, that geodesic observers are equipped with rods and clocks for measuring space-time intervals. These commoving clocks are also called "standard clocks". Standard clocks measure the proper time $d\tau = ds/c$ of the observer in question in contradistinction to the coordinate time dt, which depends on the choice of the system of coordinates. There is, however, an important restriction. Since we are discussing here cosmology within the framework of General Relativity, each geodesic observer has his own proper time, his individual time, and generally there is no means to establishing a unified time that would be relevant for all geodesic observers in the universe.

However, in cosmology we are interested in a universal cosmic time scale that allows making statements about the age and the temporal development of the universe. In order to achieve this goal we must restrict the most general problem in a convenient way. The general task to solve Einstein's field equations can be simplified considerably by introducing convenient systems of coordinates. In a first step, we make use of so-called *commoving coordinates, i.e.,* coordinates x^{μ} such that the cosmic substratum is at rest. In other words, the elements of the cosmic substratum have space-coordinates x^{k} that do not change with time. In these coordinates the line element reads:

$$ds^2 = \left(dx^0\right)^2 + 2g_{0i}\left(x^k\right)dx^0 dx^i + g_{ij}\left(x^{\mu}\right)dx^i dx^j \tag{3.1}$$

and the standard clocks of commoving observers (with constant spatial coordinates), measure coordinate time $d\tau = ds/c = dt$ and have thus the same rate. Obviously, this can always be achieved. However, the time axes of different standard clocks have still different zeros and for this reason, there is no universal concept of distant simultaneity.

In order to eliminate this ambiguity, in a second step we transform to special commoving coordinates, the time-orthogonal coordinates. In these coordinates the time axis is always orthogonal to the spatial coordinate curves and the metric assumes the very simple form:

$$ds^2 = \left(dx^0\right)^2 + g_{ij}(x^\mu)dx^i dx^j \tag{3.2}$$

In time-orthogonal coordinates we arrive at the following situation. On the world lines of the cosmic substratum clocks of proper time $d\tau = ds/c$ measure coordinate time dt and several clocks of this kind can be synchronized according to the Einstein convention by means of light rays, since in the coordinates used here the propagation of light rays is isotropic. Obviously, the time coordinate t corresponds to a universal time scale.

In time-orthogonal coordinates, Einstein's field equations can be simplified very much. The most general line element reduces to the so-called *Robertson-Walker* metric:

$$ds^2 = c^2 dt^2 - R^2(t)d\sigma^2 = c^2 dt^2 - R^2(t)\gamma_{ik}\left(x^l\right)dx^i dx^k \tag{3.3}$$

that describes in its second part the metric $d\sigma^2$ of a three-dimensional space with constant curvature – Euclidean, elliptic, or hyperbolic – together with a time dependent function $R(t)$ for the expansion (or contraction) of the substratum. This function $R(t)$ can be obtained as solution of one first-order differential equation, the *Friedmann equation* that follows from Einstein's field equations in the special situation considered here. The solutions $R = R(t)$ of the Friedmann equation provide many homogeneous and isotropic models of the universe, which can be classified according to the mass M of the universe, to the cosmological constant Λ, and to a constant $\varepsilon \in \{-1, 0, +1\}$ that determines the special kind of non-Euclidean geometry of the three-dimensional space. The function $R(t)$ describes the expansion, contraction, or pulsation of the three dimensional space. There is a huge number of exact global solutions of the Friedmann equation which correspond to models of the universe and that differ by the matter model, by invariance properties, by boundary and initial conditions, and by the large-scale topology [12].

This result confirms the so-called *Cosmological Principle* (CP). According to this principle, for any observer moving together with the cosmic substratum, the world looks homogeneous and isotropic. The two assumptions are not completely independent, since isotropy at any point implies homogeneity. The cosmological principle is not a fundamental principle but a useful working hypothesis which allows, for example, to derive the *Robertson-Walker* metric and together with Einstein's field equations also Friedmann's equation [13].

4. COSMIC TIME

On account of its hypothetical status, we will not use here the "Cosmological Principle", but instead directly establish "time orthogonal" coordinates in a more constructive and intuitive way. We consider again the simple case of incoherent matter, which is characterized by its density ρ and its velocity field u^μ. As mentioned above, we can always introduce commoving coordinates such that the matter is at rest and for a given element of the cosmic substratum the equation of motion reads $x^k = $ const. In these coordinates, the line element is given by (3.1) and the two relations $g_{00} = 1$ and $g_{0i,0} = 0$ hold. The standard clocks of the commoving, or freely falling observers measure coordinate time, *i.e.*, $ds / c = d\tau = dt$ and have thus the same rate. However, the time-axis of different standard clocks have still different zeros and can – for this reason in general not be synchronized.

The commoving character of the coordinates is not changed by the transformations:

$$x^k \rightarrow x'^k = x^k, \quad t \rightarrow t' = t + \varphi(x^i)/c \tag{4.1}$$

of the time-coordinate t with an arbitrary space-dependent function $\varphi(x^i)$. However, these transformations do change the zeros of the standard clocks by the space-dependent amount given by the function $\varphi(x^i)$. For this reason, the transformation (4.1) is a useful tool on our search for a cosmic time. Indeed, a universal concept of distant simultaneity can be achieved by an additional convention about the values of the g_{0i}'. The components g_{0i} are transformed by transformation (4.1) according to:

$$g_{0i} \rightarrow g_{0i}' = g_{0i} - \partial\varphi(x^k)/\partial x^i.$$

If we postulate that the functions $g_{0i}(x^k)$ don't change under this transformation, the transformation (4.1) reduces to:

$$x^0 \rightarrow x'^0 = x^0 + \varphi_0 \tag{4.2}$$

and shifts the zeros of all commoving standard clocks merely by the constant value φ_0, without thereby changing the distant simultaneity of the local clocks.

The most obvious convention about the components $g'_{0i}(x^k)$ of the metric tensor is $g'_{0i} = 0$. It leads to time-orthogonal coordinates with a line element (3.2) and to a time coordinate that is determined up to a transformation (4.2), which shifts the zeros of all standard clocks by the constant value φ_0. Clocks of this kind can be synchronized according to the Einsteinian convention of simultaneity, since in the time-orthogonal coordinates used here the propagation of light signals is isotropic [14].

Hence, a universal time t, that agrees with the proper time of all commoving observers, can be established, if by a transformation of the time-coordinate:

$$x^k \rightarrow x'^k = x^k, \qquad t \rightarrow t' = f^0(x^\mu)/c \tag{4.3}$$

we can achieve $g'_{0i} = 0$ for the three components of g'_{0i}. This is not always possible, but only, if certain necessary compatibility conditions are fulfilled. The physical meaning of these conditions can best be understood in terms of hydrodynamics. Indeed, they can be expressed by the requirement that the antisymmetric part $w_{\mu\nu}(x^\lambda)$ of the derivative $u_{\mu;\nu}$ of the cosmic velocity field u_μ, which describes the rotation of the cosmic fluid, must disappear [15].

This result is of far-reaching importance. It means that a universal cosmic time can be defined only if the three components g_{0i} of the metric tensor can be transformed to $g'_{0i} = 0$ by the transformation (4.3). This is, however, only possible if some necessary compatibility conditions are fulfilled. In terms of the cosmic velocity field u_μ this condition means, that the antisymmetric part $w_{\mu\nu}$ of the tensor $u_{\mu;\nu}$, *i.e.*, the rotation of the cosmic fluid must disappear. Only if there is no rotation $w_{\mu\nu}$ in the cosmic substratum, a universal time can be defined that allows for Einstein synchronization of all commoving standard clocks. Only in this case, different commoving observers that are equipped with standard clocks will agree about the numerical value of the age and the dynamical properties of the universe.

The ontological importance of this result can hardly be overestimated. In his comment to these considerations, *Max Jammer* [16] writes in this sense: *"This important result raises the critical question of whether the metric of space-time of the cosmos in which we live is really static. For it is not static, no standard synchronization can be established and Einstein's conception of time, as the equivalence class of all events induced by the standard simultaneity relation … is not merely a convention but an empty illusion".*

5. THE PARADOX OF AN INFINITE EVOLUTION

If in a cosmological model the substratum consists of incoherent matter with isotropic expansion, then we can establish a universal cosmic time which allows making statements about the age of the universe and its temporal development. Whereas for finite time intervals there are no obvious problems, new questions arise if we extend our considerations to the possibility of an infinite past of the universe. This problem has a long history that can be traced back to antiquity. We will briefly mention the main steps:

1. PLATO (427 – 347 BC.) distinguishes in his dialog *Timaios* clearly between the immutable and timeless world of the ideas that rests in eternity and the created, temporal world that is considered as an incomplete image of the world of the ideas. The image is incomplete, since time is only a moving image of eternity. As to the age of the universe there are no obvious limitations. Nevertheless, the time and the world were created together.

2. ARISTOTLE (384 – 322 BC) abandoned Plato's ontological dualism of two separated worlds, - the world of eternal ideas and the world of created and changing matter – and replaced it by one uncreated world and by an everlasting time. Hence, the Aristotelian world has no temporal limitation, it is everlasting though not eternal, since it has neither a beginning nor an end. Obviously, there is no logical need for a creator of the world.

3. PROCLUS (411 – 485), the Neoplatonic philosopher, adopted the position of Aristotle and argued in favor of an everlasting world without a beginning and without an end. His work, *On the Eternity of the World* [17], provides 18 proofs for the everlastingness of the world as arguments in favor of Neoplatonism and against the creationism of the Christians. It was subject of an extensive criticism by Johannes Philoponus, which will be discussed here more in detail.

4. JOHANNES PHILOPONUS (490 – 575) tried to disprove the 18 arguments of Proclus in favor of an everlasting world in the sense of Aristotle. In his work *Against Proclus; On the Eternity of the world* [18] he provides explicit disproves of Proclus' eighteen arguments. Before going into detail, we should clearly distinguish between the concept of eternity in the sense of Plato and the concept of everlastingness in the sense of a state lasting for ever. In accordance with Aristotle, Proclus defended the everlastingness of the world and also the objections of Philoponus are concerned with the everlasting state of the world. The main argument of Philoponus, which we will discuss here, tries to show that the age of the world must be finite and cannot be infinite as stated by Proclus. On account of its length, we will not quote the original passage in Philoponus' text but rather reformulate it in modern terms. The argument reads as follows: "*Assume, that the world has an infinite past. Then, since the beginning of the world an infinite number of successive states of the universe must have passed away. However, an infinite sequence can never be completed by a last element. Hence, in the present state, the world cannot have an infinite past*" [19].

Although the 18 arguments of Proclus have some anti-creationistic attitude, Philoponus did not consider his own arguments as a contribution to a quarrel between Christian creationism and Neoplatonic philosophy [20]. Philoponus was a Christian, but his quarrel with Proclus was on purely philosophical grounds. We add that Philoponus' refutation of the everlastingness of the world refers explicitly to the Aristotelian doctrine that the infinite cannot exist in actuality but only in potentiality. By this argument, he anticipates the modern view to infinity.

5.   BONAVENTURA (1221 – 1274); THOMAS AQUINAS (1225 – 1274). Within the framework of medieval philosophy and theology the argument of Philoponus was taken up again. Whereas Bonaventura considered the impossibility of an infinite past as an important argument in favor of the Christian creationism, Thomas Aquinas rejected any rational demonstration [21] of the finite age of the universe:

"We hold by faith alone, and it cannot be proved by demonstration, that the world did not always exist … The reason is this: the world considered in itself offers no grounds for demonstrating that it was once all new. For the principle for demonstrating an object is its definition. Now the specific nature of each and every object abstracts from the here and now, which is why universals are described as being everywhere and always. Hence it cannot be demonstrated that man or the heavens or stone did not always exist" [22].

Thomas Aquinas defended this view in a quarrel with Bonaventura at the University of Paris in about 1270.

6.   KANT (1724 – 1804). We conclude this brief historical survey with the remark that also Kant has taken up the argument of Philoponus in the *Critique of Pure Reason* of 1787. However, in the first cosmological antinomy Kant demonstrates, that Philoponus' argument against an *infinite* age of the world is a fallacy, since he can also give a (fallacious) disproof of the statement, that the world has a *finite* age. We will not go into details of Kant's argument here and refer to the literature [23].

The philosophical debate about the possibility of an infinite age of the universe has not led to a definite and convincing result. Hence, in a next step, we will ask whether the present physical cosmology can provide more information. Within the framework of Einstein's cosmology we will consider models of the universe that are homogeneous and isotropic, *i.e.,* Friedmann models. These models correspond to global solutions of Einstein's field equations. As mentioned above, in these models there are no problems to defining a cosmic time scale which allows expressing propositions about the age and the temporal development of the universe.

The solutions $R = R(t)$ of the *Friedmann equation* provides models of the universe that are homogeneous and isotropic. The solutions $R(t)$ depend on the cosmic time t, on the cosmological constant Λ, and on a constant $C = (8\pi/3)k\rho R^3 = 2kM$, that is proportional to the mass M, where k is Newton's gravitational constant and ρ the density of the mass distribution. Since the relation $M = 4\pi/3\rho(t)R^3(t)$ holds, the function $R(t)$ describes the expansion, contraction, or pulsation of the cosmic substratum. Furthermore, the solutions $R(t)$ depend on the number ε with values $(-1, 0, +1)$ that determine the non-Euclidean geometry of the three–dimensional space.

Various models of the universe can be classified according to the constants $(\Lambda, \varepsilon, C)$ with values:

$$\Lambda \in \{-, 0, +\}; \; \varepsilon \in \{-, 0, +\}; \; C \in \{+, 0\}$$

The corresponding function $R = R(t)$ can be obtained formally by integration of the *Friedmann equation*. There are many non–empty models with $C \neq 0$, $\Lambda \neq 0$ with increasing values $R(t)$, that describe an expanding universe and allow for defining the age of the world. As a special example of this kind, we mention the model $(+, 0, +)$, that can even be expressed by elementary functions.

From the beginning of the world $(t = 0, \; R = 0)$ the value of R increases with increasing t and for any finite value of R the age of the world is well defined by the inverse function $t(R)$ and is finite. (Fig. **2**). Models of this kind correspond to the widely accepted conception of the universe.

There are, however, also models with an infinite age. Of course, there is the static Einstein model $(\Lambda = \Lambda_E, +, +)$ with $\Lambda_E = 4c^4/9C^2$ and $R = R_E = 3kM/c^2$. This model does not describe an expanding universe but an unchanging, everlasting world. In addition, this solution of Einstein's field equations is unstable. It decays into the Eddington-Lemaître solution $(\Lambda = \Lambda_E, +, +)$ with increasing $R = R(t) > R_E$.

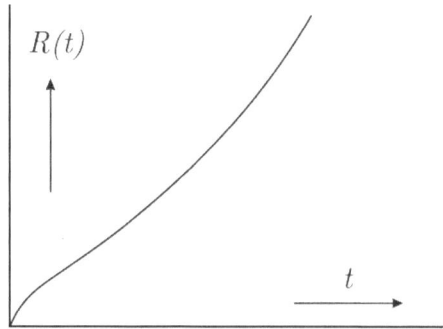

Figure 2: Graphical representation of the model $(+, 0, +)$. Expansion Parameter $R(t)$, cosmic time t.

The Eddington-Lemaître model is of particular interest for our problem, since in the limit $t \to -\infty$ the function $R(t)$ decreases from the present finite value $R(0) > R_E$ to $R = R_E$, *i.e.,* $\lim\limits_{t \to -\infty} R(t) = R_E$. For large values of $t > 0$ the function $R(t)$ increases as $R(t) \approx \exp\{ct(\Lambda_E/3)^{1/2}\}$; (Fig. **3**).

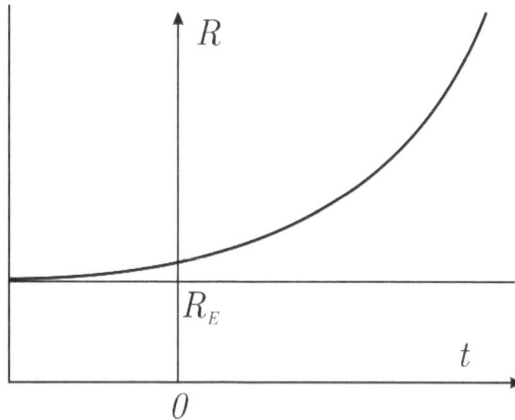

Figure 3: Graphical representation of the Eddington-Lemaître universe. Expansion parameter $R(t)$, cosmic time t, radius R_E of the Einstein universe.

The Eddington–Lemaître model describes a universe with infinite age. Its beginning is in the infinite past, at $t \to -\infty$ with the initial value $R = R_E$, and it is expanding since that time.

$R_0 = R(0)$ corresponds to the present value of R. Obviously, there are no inconsistencies in this way of reasoning. The philosophical objections against the infinite past, *i.e.,* the argument of Philoponus, shows merely how we must not speak about an infinite age of the universe. The present time must not be identified with the last element of an infinite sequence of steps.

6. ETERNITY – BEING WITHOUT TIME

For an explication of the concept of eternity that is for the present free from theological connotations we go back again to PLATO and refer as above to his dialog *Timaios*. In this dialogue "eternity" is an attribute of the immutable and timeless world of ideas, which is ontologically prior to the world that is created as similar as possible to the world of the ideas by the creator of the world, the Demiurge. There is, however, an insurmountable difficulty, which comes from the dissimilarity of the two worlds:

> *"Now the nature of the ideal was everlasting, but to bestow this attribute in its fullness upon a creature was impossible. Wherefore he resolved to have a moving image of eternity, and when he set in order the heaven, he made this image eternal but moving according to number, while eternity rests in unity; and this image we call time".* *(Timaios 37B)*

According to these conceptions of time and eternity, Plato writes that:

> *"Time and the heaven (the world) came into being together in order that, having been created simultaneously, if ever there was to be a dissolution of them, they may be dissolved simultaneously".(Timaios 38B)*

It is obvious, that the formulation "time and the heaven came into being together" could lead to logical problems, if the word "together" is understood as "simultaneous", *i.e.,* in the temporal interpretation and not in the etymological one, that understands ἅμα in the neutral sense of "together" [24]. That Plato used "together" in the temporal sense is confirmed by another passage preceding the quoted statement in *Timaios* 38B: *"there were no days and nights and months and years before the heaven was created".* (*Timaios* 37C)

Since according to the same text (*Timaios* 38B), time and the moving planets were created together, some authors argued, that Plato had identified time with the motion of planets. However, in the *Timaios* we cannot find such identification. Indeed, Plato calls the planets also "tools of time" since they allow to measure time quantitatively.

The logical problem of a simultaneous creation of time and the world was discussed since the days of Plato by many scholars, from Aristotle through Augustinus to physicists working in modern cosmology. We mention here in particular the discussion of these problems by AURELIUS AUGUSTINUS (354–430) in "*Confessiones*" and in "*De civitate dei*" [25, 26]. Augustinus discussed in detail the logical problems that arise, if we assume that God created not only the world but together with the world also the time. In this way, he is led to the questions like: "*What has God done before He made heaven and earth?*" [27] The answer which was given by Augustinus in his "*Confessiones*" leads to further clarification. He claimed that before God created heaven and earth, he made nothing. For, if he had made something, it would have been nothing else but a creature and Augustinus was deeply convinced that no creature came into being before the world was created. And he added: "*If before heaven and earth there was no time, why is it demanded, that Thou then didst? For there was no 'then', when there was no time*".

An additional reason for this answer is elaborated more in detail in "*De civitate dei*". Here, Augustinus argued, that time is not only the framework in which the world was created, but that time could not come into being without entities of the material world that allow to observe and to measure time intervals and that for this reason time and the world must have been created together. "*Who does not see that time would not have existed had not some creature been made, which by some motion would bring about change, and that since the various parts of this motion and change cannot exist simultaneously when one passes away and another succeeds it in shorter or longer intervals of duration, time would be the result?*" [26]. Obviously, in this passage Augustinus anticipates in some sense the operational definition of time of the 19th and 20th century.

Another interesting facet of the concept of eternity is discussed in the *Consolationes* of BOETHIUS (480-524) who connects aspects of ancient philosophy with new thoughts of the Christian theology. In accordance with Aristotle he assumes the temporal infinity of the world in the sense that there is no beginning and no end. In contrast to the semi-eternity of the created world, eternity, as an attribute of God, is understood as "the complete simultaneous and perfect possession of interminable life". ("*aeternitatis est interminabilis vitae tota simul et perfecta possessio*") [27]. Eternity is not the everlastingness of the created world, which was assumed by Aristotle, but it comprehends at once the successive temporal stages of past, presence, and future.

Boethius argues, that this understanding of eternity allows for an important distinction of two different kinds of recognition of the world by an eternal being and a human one. Whereas a human being can recognize at most the past and the presence, but not the indeterminate future, God knows at once, simultaneously, together, past, presence and future. Future events, which are considered in a human perspective partly as a result of our free will, appear in the eternal perspective of God as necessary. Hence, predictability and necessity of future events depend generally on whether the perspective is eternal or human. Obviously, we are confronted here with the intricate problem, whether determinism of natural events, free will of human beings, and divine ubiquity are compatible. Boethius defends the consistent compatibility of these aspects. He assumes determinism in the sense of Aristotle [28], he considers free will as indispensable for any rational being, and he considers *"omniscience"* as a necessary predicate of God. The compatibility of human free will and divine omniscience results from the incompatibility of the completely different kinds of recognizing the world by human beings and by God. For this reason, human beings cannot change by their free will events that are providential. ("*divinam te praescientiam non posse vitare, …*" [29]). We will not try to reformulate here this argument such that it fulfils modern standards of exactness and conclusiveness. Instead, we will rather investigate the question, whether parts of this way of reasoning can be reconstructed in terms of a modern scientific theory of space-time.

General Relativity is the widely accepted theory of space, time, and gravity in modern physics. As mentioned in the preceding sections, this theory is governed by Einstein's field equations, which connect the energy momentum tensor $T^{\mu\nu}$

with the geometry of a four dimensional manifold $\mathcal{M}$. For the cosmological problems discussed here, the global solutions of Einstein's field equations are of particular interest, since they may be considered as models of the universe. In general, global solutions are infinity extended in space and time. Observers, who describe the space-time manifold by observations and measurements, are connected with a reference frame that is formally represented by a velocity field.

A model of the universe is given by a four-dimensional manifold $\mathcal{M}$ and a metric g of signature 2. In many cases we can characterize space-time models by exact solutions of Einstein's field equations. A local observer describes the system $(\mathcal{M}, g)$ on the basis of a reference frame and by means of convenient coordinates, – thus describing the world from inside. This is the view that provides our usual perspective of the sky, planets, stars, galaxies *etc.* However, the possibilities of local observers to gain information about the structure of the entire space-time $(\mathcal{M}, g)$ is restricted by the large scale topology and by the light cone structure of the universe. The observer is moving on a time-like worldline and his knowledge about the world is restricted to well defined regions of space-time. Even if an observer could observe the world up to the end of time, his information will in general not be complete, since in many global solutions there is no "doomsday". This is, for example, the case in the empty de Sitter universe. We will not go into details here.

However, we can also leave this local perspective of the universe and replace it by a global view. The global perspective that we have in mind is rather a bird's eye view of the temporal developments in the world. Of course, there are no longer observers located in finite regions of space-time. Formally, the transition from the local perspective to the global view can be described by a sequence of coordinate transformations and conformal transformations of the metric. For sake of simplicity, we illustrate this procedure at the Minkowskian space-time [30].

In this way, we arrive at a completely finite picture of a world $(\mathcal{M}, g)$ that is infinitely extended in space and time. This "Penrose diagram" of the world $(\mathcal{M}, g)$ is the finite and complete grasp of a world that is interminable in space and time. The new coordinates (r', t') provide a picture of the universe from outside that is immutable, since all temporal developments within the world are already contained in this picture. The time-like world-lines are completely shown in this diagram, from the infinite past to the infinite future. Hence, we could consider the Penrose-diagram as the picture of the world as seen from an eternal perspective. If the analogy is not taken too literally, the philosophical picture of world and eternity that was conceived by Boethius can be reconstructed in terms of modern physics by Penrose-diagrams of the universe. (Fig. **4**).

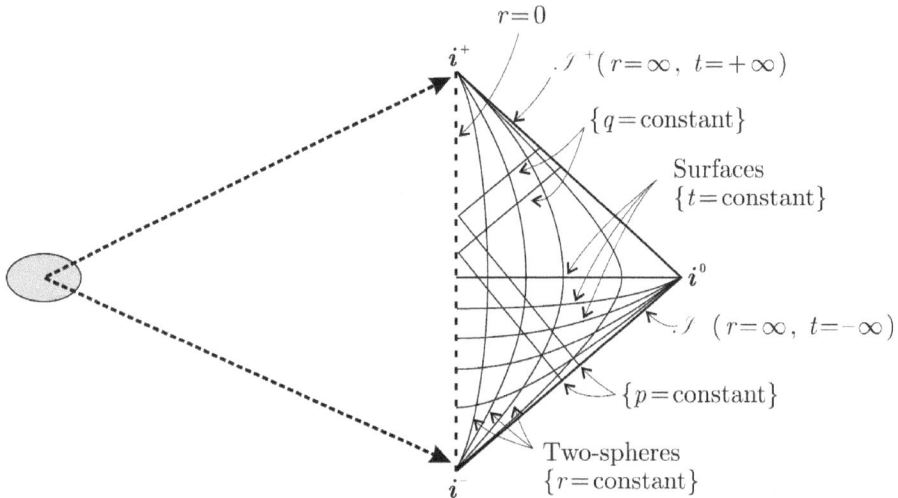

Figure 4: Penrose diagram of the Minkowskian space-time (M, g). The entire world is represented by the triangle; the observer describes the world (sub specie aeternitatis) from outside. Each point in the triangle represents a two-sphere except for the single points i^+, i^0, and i^-, and the points on the line $r = 0$.

Any time-like geodesic is starting at i^- (infinite past) and finishing at i^+ (infinite future).

7. CONCLUSIONS

Within the framework of General Relativity, a universal cosmic time that is relevant for all observers who are commoving with the cosmic substratum can be established by a convenient system of coordinates only, if the substratum is free from vortices. Hence, the existence of a cosmic time depends on contingent properties of the cosmic substratum. – Only if these properties are given, the age and the dynamical behavior of the universe can consistently be described. Under this assumption, we discussed the question, whether the universe could have an infinite past and showed that – in spite of a long lasting philosophical debate -

there are consistent physical models with infinite age. Furthermore, we argued that the concept of eternity as it was conceived in the philosophical tradition since the days of Plato, can be given an adequate meaning within the context of modern relativistic cosmology.

ACKNOWLEDGEMENTS

Declared none.

CONFLICT OF INTEREST

The author(s) confirm that this chapter content has no conflict of interest.

REFERENCES

[1]   Newton, I. (1687), Philosophia Naturalis Principia Mathematica, London, Streator.
[2]   Newton, I. (1934), Mathematical Principles of Natural Philosophy, F. Cajori (ed.), University of California Press, Berkeley.
[3]   Newton, I. (1934), l.c.
[4]   Jammer, M. (2006), Concepts of Simultaneity, The John Hopkins University Press, Baltimore, p. 73.
[5]   Mach, E. (1901), Die Mechanik in ihrer Entwicklung, 4. ed., Leipzig, F. A. Brockhaus. English translation: "The Science of Mechanics", (1902), (transl. by T. J. McCormack) Chicago.
[6]   Poincaré, H. (1898), La mesure du temps, Revue de metaphysique et de moral, t.VI, (janvier 1898) pp. 1-13.
[7]   The problem of distant simultaneity is considered in great detail systematically as well as historically in the book of M. Jammer, Jammer (2006), l.c.
[8]   The conventionality of distant simultaneity cannot be avoided by directly measuring the one-way velocity of light, since this is not a measurable quantity. Cf. Jammer (2006), l.c. p. 232 ff. and the literature quoted there.
[9]   For mathematical details we refer to Misner, Ch. W., Thorne, K. S., and Wheeler, J. A. (1973) Gravitation, W. H. Freeman, San Francisco; and to Rindler, W. (1977) Essential Relativity, Springer, Heidelberg.
[10]   Heintzmann, H. und P. Mittelstaedt, (1968), Physikalische Gesetze in beschleunigten Bezugssystemen, Springer Tracts in Modern Physics, **47**, (1968), pp. 187-225.; Mittelstaedt, P. (1989), Der Zeitbegriff in der Physik 3.ed. BI-Wissenschaftsverlag, Mannheim, 1989, Ch. V.
[11]   Generally, for the 4 indices of space-time coordinates we use Greek letters, for the 3 indices of space-coordinates Latin letters. The coordinate $x^0 = ct$ denotes the coordinate time t.
[12]   Examples for exact global solutions can be found in many monographs, *e.g.,* : Hawking, S. W. and G. F. R. Ellis, (1973), The large scale structure of space-time, Cambridge University Press, Chapter V.

[13]   Cf. Rindler, W. (1977), Essential Relativity, Springer-Verlag, Heidelberg, pp. 210-211. See also the discussion of the "Cosmological Principle" in the contribution of K. Mainzer in this volume.

[14]   Jammer, M. (2006), Concepts of Simultaneity, The John Hopkins University Press, Baltimore, p.280.

[15]   Mittelstaedt, P. (1989), Der Zeitbegriff in der Physik, BI-Wissenschaftsverlag, Mannheim, pp. 156-165.

[16]   Jammer, M. (2006), l.c., p.280.

[17]   Proclus (2001), On the eternity of the world, Helen S. Lang and A.D. Macro (eds),, Berkeley.

[18]   Philoponus (2004), Against Proclus; On the Eternity of the World, Transl. Michael Share, London, Duckworth.

[19]   Johannes Philoponus, l.c. Chapter 1, section 3, no.9, (pp. 23 – 24)

[20]   For this interpretation of the work of Philoponus l.c., we refer to the introduction of this volume by M. Share.

[21]   Observe that by "demonstration" Thomas Aquinas means a rigorous proof from premises which essentially include necessary laws (of nature).

[22]   Thomas Aquinas (1948), Summa Theologica, I, 46, 2; Translated by Fathers of the English Dominican Province, Westminster, Maryland.– For details cf. Mittelstaedt, P. and P. Weingartner (2005), Laws of Nature, Heidelberg, Springer, pp. 70 and 284-295.

[23]   Mittelstaedt, P. and I. Strohmeyer, (1990) Die kosmologischen Antinomien in der Kritik der reinen Vernunft und die moderne physikalische Kosmologie, Kantstudien 81, pp. 145 – 169; Malzkorn, W. (1999), Kants Kosmologie-Kritik, Berlin, Walter de Gruyter.

[24]   Jammer (2006), l.c., p. 31

[25]   Aurelius Augustinus, Confessiones, book 11, chapter 12-13.

[26]   Aurelius Augustinus, De civitate dei, book XI, chapter 6.

[27]   Boethius, Consolationes, book V, chp. 6, 9-10.

[28]   Aristotle, Physics II. 4-6.

[29]   Boethius, Consolations, book V, chp. 6, 150.

[30]   For more details cf. Hawking, S. W. and G. F. R. Ellis, l.c. pp. 118-124.

Send Orders of Reprints at reprints@benthamscince.net

CHAPTER 4

Objective and Subjective Time in Anthropic Reasoning

Brandon Carter[*]

CNRS, LuTh, Observatoire Paris – Meudon, France

Abstract: The original formulation of the (weak) anthropic principle was prompted by a question about objective time at a macroscopic level, namely the age of the universe when "anthropic" observers such as ourselves would be most likely to emerge. Theoretical interpretation of what one observes requires the theory to indicate what is expected, which will commonly depend on where, and particularly when, the observation can be expected to occur. In response to the question of where and when, the original version of the anthropic principle proposed an *a priori* probability weighting proportional to the number of "anthropic" observers present. More refined versions would adjust this by an anthropic quotient allowing for the relative rate of subjective (sentient) mental processing (which would presumably have been lower in extinct hominids than in modern humans). The present discussion takes up the question of the time unit characterising the biological clock controlling our subjective internal time, using a revised alternative to a line of argument due to Press, who postulated that animal size is limited by the brittleness of bone. On the basis of a static support condition depending on the tensile strength of flesh rather than bone, it is reasoned here that our size should be subject to a limit inversely proportional to the surface gravitation field g, which is itself found to be proportional (with a factor given by the 5/2 power of the fine structure constant) to the gravitational coupling constant. This provides an animal size limit that will in all cases be of the order of a thousandth of the maximum mountain height, which will itself be of the order of a thousandth of the planetary radius. The upshot, *via* the (strong) anthropic principle, is that the need for brains, and therefore planets, that are large in terms of baryon number may be what explains the weakness of gravity relative to electromagnetism.

Keywords: Biological evolution, anthropic principle, multiverse, gravitational attraction, doomsday scenario, environmental catastrophe, hyperbolic population growth, weakness of flesh, planetary radius, timescale of perception.

1. INTRODUCTION

In this interdisciplinary forum for the discussion of various kinds of time, one of

*Address correspondence to Brandon Carter: Emeritus Director of Research, CNRS (National Centre for Scientific Research), Observatoire de Paris, 92195 Meudon, France; Tel: +33 1 45344677, 45077434; Fax: +33 1 4507 7971; E-mail: brandon.carter@obspm.fr

Argyris Nicolaidis and Wolfgang Achtner (Eds)

the first questions that comes to mind is that of the relationship between external time of the kind measured by physicists in the objective world, and the internal psychological time characterising successive instants of conscious perception by a human (or other comparable) mind. As this is something I had previously thought about in the context [1] of anthropic reasoning for astronomical and other purposes (notably in the development [2, 3] of a microanthropic principle such as seems necessary for the interpretation of quantum theory) my objective here will be to offer a brief account of what emerges from that point of view.

Before recapitulating what is meant by the notion of anthropic reasoning, I would start by explaining that this essay will take for granted the usual paradigm whereby it is assumed that the mind, as perceived from inside, corresponds, in a material physical world, to an organism of the kind known as a brain. More specifically, the motivation for the following discussion derives from the supposition (invoked in previous relevant work by Dyson [4] and Page [5]) that mental processes are describable in terms of instants of perception characterised by a finite duration,

$$\tilde{\tau} = \delta t \tag{1}$$

of time t – interpretable as a basic biological clock unit – as measured in the physical world wherein the brain is situated. In the human case it would seem that the shortest biological clock timescale on which coherent mental processes occur is of the order of a fraction of a second, which is of course why the latter has been chosen as the standard time measurement unit.

It has for more than a century been possible, and for more than half a century been common usage, to work, as will be done here, with what are known as Planck units. In this (unrationalised) system the fundamental physical measurement units of mass, of length, and (for our present purpose most pertinent) time are chosen so as to give unit value to three particularly important fundamental parameters. The first of these is the coupling constant G that was crucial for what was historically the earliest satisfactory physical theory namely Newtonian gravitation. The second is the propagation speed c that was crucial for the next great physical theory, namely that of Maxwellian electromagnetism, which was subsequently unified with gravitation by Einstein. The third is the more mysterious parameter designated $\hbar$, of which the importance was first recognised by Planck, but of which the significance was not properly understood until (synthesising contributions by Schrödinger, Heisenberg, Born and others) the principles of modern quantum theory were finally established by Dirac.

An immediate consequence of this astounding breakthrough was to provide an explanation of the complexity of the chemical behaviour of the elements, as described by the Mendeleev's puzzling periodic table, in terms of just a single dimensionless coupling constant, namely the charge of the electron, whose square, the fine structure constant $\alpha_e = e^2$, is given approximately, in these units, by $e^2 \doteq 1/137$, while a complete account of low energy physics on a local state required only one more dimensionless parameter namely the electron proton mass ratio:

$$m_e/m_p \doteq 1/1830$$

which which is important for the properties of liquid helium. These values, and those of the Yukawa coupling constant $a_Y \doteq 2/7$ and the pion proton mass ratio $m_\pi/m_p \doteq 1/7$ that play an analogous role in the much less complete theory of strong interactions introduced about the same time, were fundamental in the sense of being obtainable only by empirical observation. Early hopes that they would soon be obtainable by calculation from some deeper theory remain unfulfilled half a century later.

Although there has of course been progress in the derivation of Yukawa's rudimentary pion coupling model from more sophisticated strong interaction theories involving quarks and gluons, this has been done only by introducing even more adjustable parameters than before, so the outcome is that the value of the effective coupling constant a_Y still remains something known only by empirical measurement. Therefore, to account for the value of this and other such quantities, there is now more interest than ever in what I called the strong anthropic principle [6]. This means an approach whereby the quantities in question are postulated to have values that vary over observationally inaccessible parts of what recently come to be known [7] as a multiverse, within which our own region has been selected by the existence of – and therefore environmental suitability for – observers like ourselves.

It would appear, as I have argued elsewhere [8], that in order to make sense of the Everett approach to quantum theory it is essential in any case to invoke the concept of a multiverse as an arena consisting of channels subject to selection by application of the anthropic principle in its ordinary "weak" form. (According to this anthropic implementation of the Everett interpretation, the evolution of the quantum state of the multiverse is unitary, and thus in principle deterministic, but in practice one is nevertheless subject to effective uncertainty because one does not know in advance which branch one will turn out to be on). The more particular "strong" proposal that such a multiverse should provide a range of different possibilities for the values of fundamental paramenters was not widely welcomed when I originally suggested it [6], but this idea has come to be taken more seriously in recent years, due to the development of superstring theory, which requires the invocation of a "landscape" [9-11] that does indeed involve the requisite range of possible parameter values.

As the prototype candidate for such a "strong" anthropic selection effect, I had drawn attention [12] to the observed Yukawa coupling relation:

$$a_Y \doteq 2m_\pi/m_p, \tag{2}$$

that is well known as the condition for the nuclear coupling to be marginally strong enough for the formation of a deuteron by binding of a proton and a neutron within the distance fixed by the pion mass. As consequently remarked by Dyson [13] and confirmed by subsequent calculations [14], a relatively small respective decrease or increase in this coupling would suffice to provide a chemically sterile universe consisting exclusively of hydrogen, or containing none at all.

Whereas most such fundamental coupling constants were found to have values more or less comparable with unity, Dirac was impressed by the fact that there is an exception. Compared with their electromagnetic attraction, the gravitational attraction between an electron and a proton is weaker by an enormous factor of the order of 10^{40}, a number that is interpretable as roughly the inverse of the product of the mass of these particles, as given in Planck units roughly by $m_e \approx 10^{-22}$ and $m_p \approx 10^{-19}$, so that their geometric mean square value is given in very round figures by:

$$m_e \, m_p \approx \langle m \rangle^2 \approx 10^{-40} . \tag{3}$$

The point to which I wish to draw attention here – and for which at the end I shall offer a tentative explanation – is that in terms of such Planck units the fraction of a second time unit characterising our mental (and other biological) processes has a value of about the same enormous order of magnitude, namely:

$$\tilde{\tau} = 10^{40} . \tag{4}$$

2. DIRAC'S COINCIDENCE

Dirac himself drew attention, not to the coincidence I have just mentioned,, but to another coincidence involving the same number, which is that it is roughly the square root of the number $N \approx 10^{80}$ of protons (or cquivalently, by charge neutrality, of electrons) in the visible universe, meaning the volume characterised by the cosmological Hubble radius, which was first measured (though not very accurately) about the same time (three quarters of a century ago) that the principles of modern quantum theory were first established. It was Dirac's mistaken explanation for this highly significant coincidence that prompted me to provide an explicit formulation of what I called the anthropic principle.

Dirac's idea was based on the supposition that we are observing at a random – so presumably typical – instant in the history of our expanding universe, so if (for whatever reason) the inverse of the gravitational coupling constant is presently equal to the square root of the number of particles in the visible volume determined by the Hubble expansion rate then it should be expected to remain so. Since the visible volume will include progressively more and more particles as the universe expands, it would follow that the gravitational coupling should become correspondingly weaker.

When I first read about this (in Bondi's classic textbook [15] on cosmology) I realised that there was a weak link in Dirac's reasoning, namely his implicit adoption of what I would criticise as a ubiquity principle rather than the anthropic principle that seems appropriate. The essential content of what I called the anthropic principle is that, *a priori*, our location in space or time should not be considered as random with just with respect to the corresponding ordinary physical measure of space or time (as in what I would call an ubiquity principle) but with respect to a measure that is anthropically weighted in the sense of being proportional to the population density of individuals comparable with ourselves. In a more precise treatment, this factor would involve an anthropic quotient that would presumably be less than unity for exinct hominids [16].

It has since become observationally clear that the conclusion to which Dirac was drawn by his misguided line of reasoning was indeed wrong, since his predicted (cosmological rapid) weakening of gravitation does not actually occur. Already, even before this was as obvious as it is now, the weak point in Dirac's line of reasoning had been noticed by Dicke [17] (an advocate of another theory of progressive weakening of gravity, but at a much slower and so less easily measurable rate) who preceded my own contribution [5] in pointing out that observers comparable to ourselves (which is what I meant by the adjective 'anthropic') could not have existed when the age of our expanding universe was much less than the lifetime of a typical hydrogen burning star, and could be expected to become relatively rare when the universe is much older that that. The (then recently discovered) reason for the lower cut off is that such hydrogen burning is the only way of fabricating the medium and heavy elements of which we are made. The (more obvious) reason for the upper time limitation on the relevant anthropic weighting measure is the dependence of life systems such as ours on energy input from a neighbouring star, and of the fact that although later generations of such stars will continue to be formed they will become progressively rarer as the matter that was originally present is transformed into unavailable end states such as cold dead neutron stars and black holes.

Although Dirac was wrong in his conclusion of weakening of gravitation with time, it appears that he was right in deducing that there is a direct connection between the strength of gravity and the number of particles in the universe as presently observed. The coincidence he noted is that the value, roughly 10^{80}, of N is roughly the inverse square of the value, roughly 10^{-40}, of the gravitational coupling constant that is itself given very roughly as the square of the proton mass m_p. Dirac's coincidence is thus roughly expressible in Planck units simply as:

$$N \approx m_p^{-4}. \hspace{4cm} (5)$$

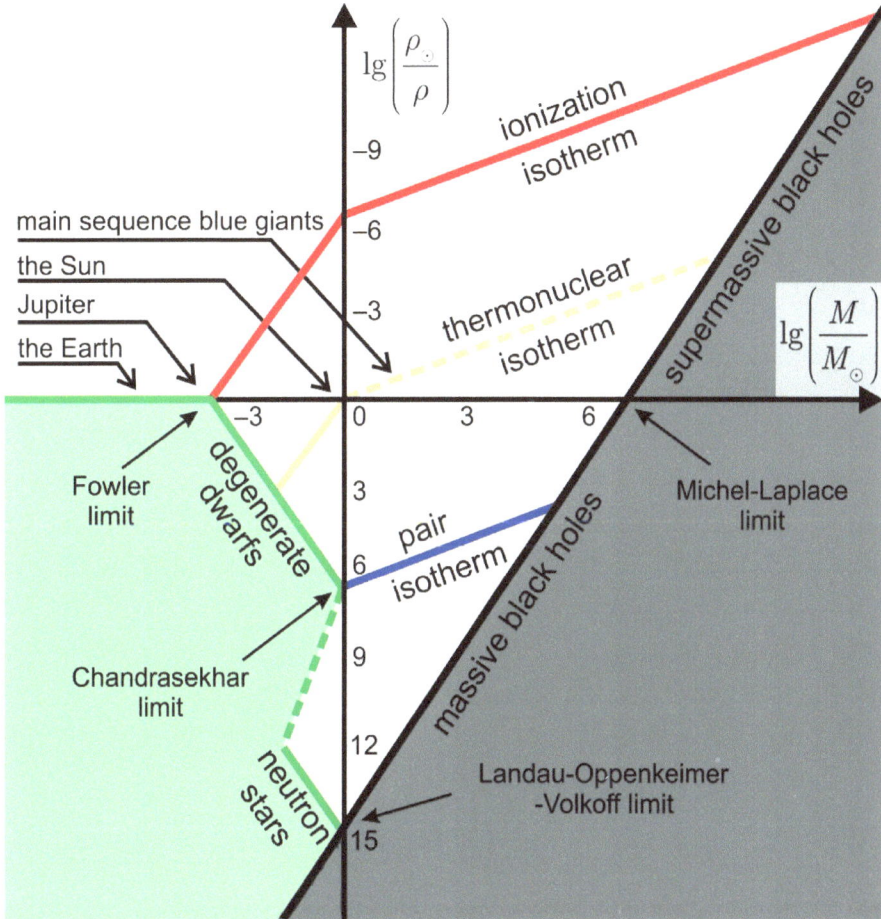

Figure 1: Logarithmic plot of inverse density against mass for the main kinds of stellar and planetary bodies, as accounted for [9] (on the basis of work by pioneers such as Eddington and Chandrasekhar) in terms just of the masses and charge of the proton and the electron.

A first step in what now appears to be the correct explanation of this is the recognition that, as has been known since Newton's time, the dynamical timescale t associated with the free self gravitational acceleration of a body with mass density ρ will (remembering our units here are such that G = 1) be given roughly by $\rho \approx 1/t^2$. In the cosmological case (remembering our units are such that $c = 1$) the relevant Hubble radius will be of the same order as the age t, so the order of magnitude of the mass in the visible volume will be given roughly by $M \approx \rho t^3 \approx t$ which means that the corresponding number N of particles with mass m_p will be given by:

$$N \approx t / m_{\mathrm{p}} \quad .$$

Dirac's coincidence is therefore equivalent to the (directly verifiable) observation that the present age of the universe is given very roughly by:

$$t \approx m_{\mathrm{p}}^{-3}, \tag{6}$$

which in very round figures comes out to be something like 10^{60}.

3. EXPLANATION FROM STELLAR PHYSICS

To explain Dirac's coincidence on the basis of the anthropic line of reasoning that was implicitly followed by Dicke, it suffices to work out the typical lifetime τ_* of a main sequence star, something that was first understood, on the basis of work by Eddington and others, about the same time as Dirac was clarifying the essential principles of quantum theory. Since a star is held together by gravity, it is not surprising (see my recent recapitulation [18]) that its properties should be essentially determined by the gravitational coupling strength: the essential conclusions (see Fig. **1**) are that typically (give or take one or two factors of ten) the mass M_* will be given roughly by:

$$M_* \approx m_{\mathrm{p}}^{-2},$$

while its lifetime, which is what we are principally concerned with here, will be given by $\tau^* \approx \varepsilon_{\mathrm{N}} \, M_*/L$ where L is its luminosity, and ε_{N} is the nuclear binding energy which, on account of (1), will be expressible simply as $\varepsilon_{\mathrm{N}} \approx (2m_\pi/m_{\mathrm{p}})^2$. The luminosity is sensitive to the stellar mass, and is given by a rather complicated formula [12] for small slow burning stars. However, for larger brighter stars of the kind that manufactured our heavy elements, one can use the simple Eddington luminosity formula $L/M_* \approx m_{\mathrm{p}} \, m_e^2/e^4$ which gives

$$\tau_* \approx (2 e^2 \, m_\pi / m_e)^2 \, m_{\mathrm{p}}^{-3}. \tag{7}$$

Since (in our part of the universe, if not elsewhere) the factor $e^2 m_\pi/m_e$ happens to be of order unity, it is evident that (at the admittedly rather crude level of accuracy involved) the agreement with (5) is perfect: the coincidence is explained, and the essential role of gravity confirmed. However, instead of providing evidence for a radically new theory as foreseen by Dirac, the explanation merely confirms what was already the established understanding of the situation. Dirac's persistent refusal to accept this (from his point of view disappointing) outcome shows that, despite the fact that it ultimately contributes nothing new in such a case, the anthropic principle is not merely a tautology. Indeed, whereas Dirac's inclination was to spread the a priori probability measure too widely, a more common kind of deviation from the anthropic principle is to spread it too restrictively. Both kinds of deviation tend to be motivated by wishful thinking, typically unwillingness to accept the limitations, particularly concerning future prospects, that are involved in unbiased application of the anthropic principle.

4. ANTHROPIC OR AUTOCENTRIC FORECAST

An extreme, but not unusual, example of a non-anthropic principle of the restrictive kind is what I would call the autocentric principle, whose application consists in a (logically admissible, but for predictive purposes sterile) refusal to attribute any a priori probability at all to situations other than that in which one has already found oneself *a posteriori*. (As a recent proponent [19] puts it, "what other hypothetical observers with data different from ours might see, how many of them there are, and what properties they might or might not share with us are irrelevant").

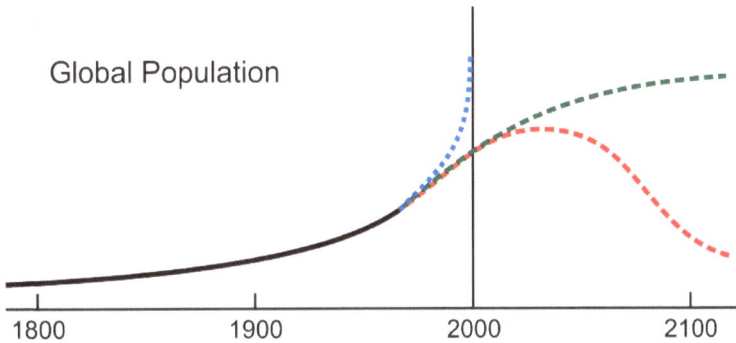

Figure 2: Future of terrestrial human population, as envisaged circa 1980. A deviation from the high scenario of continually accelerating growth is already be-coming observable. If this tendency is confirmed, the remaining possibilities are the medium scenario of sigmoid growth toward saturation at "sustainable" level – (which is however unlikely according to anthropic reasoning) and the low scenario of quasi-normal evolution (which is most realistic inview of the exhaustion of non-renewable resources).

Recourse to such an autocentric standpoint may be a temptation for advocates [20] of scenarios in which the population of our terrestrial civilisation is maintained indefinitely at a sustainable level (as envisaged by ecologists) or even conceived to undergo continued growth (as desired by economists). Whereas the autocentric principle is compatible with (neither encouraging nor discouraging) such scenarios, on the other hand, according to the anthropic principle, the integrated future population of our civilisation can not be expected to greatly exceed the number that have lived so far, which leaves us with the choice between a sudden cut off (what has been called the doomsday scenario [21], due to a catastrophe like a global war or some kind of environmental disaster) or the less apocalyptic alternative of a gentle and controlled decline in population, of the kind that seems already to have begun in some developed countries. In the interval since I first made such a prediction nearly thirty years ago [22], official United Nations statistics indicate (see Fig. **2**) that there has already been an inflection, whereby, at the very end of the 20th century, the second time derivative of the globalhuman population changed sign and became negative, thus terminating a long period in which, for recently elucidated reasons [16], growth was not just exponential but hyperbolic. Assuming it is not a just short term fluctuation, this inflection is in line with what is to be expected anyway, on the basis of environmental and other considerations, not just on the basis of the anthropic prediction. The implications of the latter can therefore not be easily written off, even by recourse to autocentrism.

5. UNDERSTANDING OUR PAST EVOLUTION

Although it is perhaps of less immediate practical importance, a more intellectually interesting application [22] of the anthropic principle is concerned with another remarkable coincidence concerning the stellar lifetime discussed above, and in particular with the more precise value that can be given to it when one is concerned not with the whole category of main sequence stars but with a single specific case, namely that of our own Sun. The remarkable coincidence (much more precise than the vague order of magnitude relation that fascinated Dirac) is that the predicted total hydrogen burning lifetime of the Sun is only about twice the present geologically measured age of our planet Earth. What this means is that the stochastic biological evolution process whereby our (at one stage single celled, then worm like, fish like, and finally mammalian) ancestors developed to become our anthropoid selves took fully half the astrophysically available time. If this very complicated and technically (unlike stellar astrophysics) not at all well understood process had been slower by a modest factor of two our civilisation here would never have got to exist at all. It is not at all plausible that the intrinsic stochastic mechanism of such a biological process should have been tuned a priori so as to agree with the externally imposed astrophysical timescale (see Fig. **3**).

What could have been expected a priori is that in the given planetary environment, the stochastically expected time scale, $\bar{t}$ say, for evolution to our anthropically advanced state would be either very short or else very long compared with the relevant stellar lifetime. In the former case advanced civilisations would be of rather common occurrence and the anthropic weighting factor would favour their emergence when the relevant star was relatively young. Since that is not what we observe in our own case, we are left, as the only plausible alternative, with the conclusion that the available time is short compared with the stochastically expected timescale $\bar{t}$, which implies that (whereas primitive life may be relatively common [23]) advanced civilisations like ours will occur only in rare cases for which there was an exceptional run of luck. More detailed conclusions [24, 25] can be obtained by classifying the intermediate evolutionary steps as either easy ones, meaning those with a high chance of going through in the time available, or difficult ones, meaning those with a low chance of going through in the time available. Our observation that the Sun is no longer young implies that at least one of the steps must have been of the difficult kind, but on the other hand the number of difficult steps can not have been very large since if it had the remaining main sequence time would have been expected to be much smaller than, not comparable with, the time elapsed so far.

6. OUR PLANETARY ENVIRONMENT

The smallest timescales of which we have any actual experimental or observational knowledge are those of nuclear physical processes for which the relevant timescales are of the order of m_{p}^{-1} meaning about 10^{20} in Planck units. The very long cosmological, stellar, and biological evolution timescales whose anthropic relationships have been discussed are of the order of m_{p}^{-3}, meaning about 10^{60} in Planck units. Between these very large and very small values, the minimum (fraction of a second) perception timescale $\tilde{\tau}$ referred to in (3) is given by the geometric mean, namely about 10^{40} in Planck units. As this happens to be what is recognisably expressible as m_{p}^{-2} we again find ourselves confronted with the question of whether this is just an accidental coincidence, or whether, as with Dirac's coincidence, it really is explicable in terms of gravitational coupling.

A significant step toward such an explanation is contained in the pioneering investigation of the physical and astrophysical limitations on human space dimensions provided [26] by Press, whose key point was that the dimensions of the host planet are rather tightly restricted by the requirement that the gravitational field should strong enough for binding of water and heavier molecules, but not quite strong enough for binding of hydrogen. This means that, compared with the square of the escape velocity, the thermal energy factor given for the atmospheric nitrogen and oxygen by the square of the sound speed has to be rather (but not too) small – by a factor of roughly the order of a thousandth.

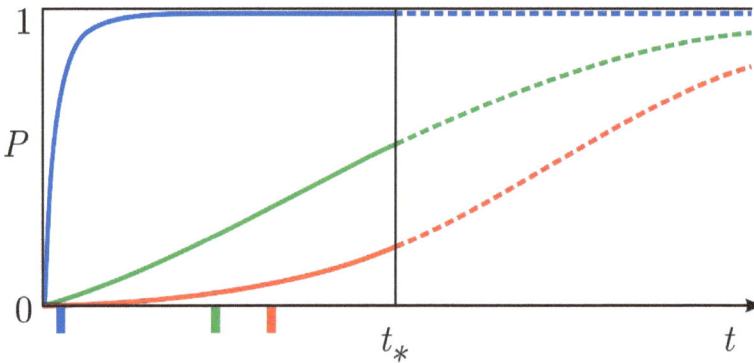

Figure 3: Probability distribution and expectation value for anthropic arrival time, firstly if the number of difficult steps is zero (high curve), secondly if it is one (middle curve) and thirdly if it is two (low curve), using dotted lines for extrapolation beyond the cut-off imposed by the Sun's lifetime. The expectation value in the first case is far too low to account for the observation that the present age of the earth is about half the calculated cut-off time. However the second case, that of a single difficult step, fits very well, and it is still possible to envisage that there may have been two difficult steps (or even more if – to allow for rising of the solar temperature – the standard cut-off calculation needs downward revision).

For marginal gravitational binding of hydrogen atoms (with thermal velocity not far below the escape velocity) the generic virial equilibrium formula – as explained in my recent recapitulation [18], for a planetary or non-relativistic stellar type body of mass M, density ρ and temperature Θ, takes a form given roughly by $M \approx 10\,m_p^{-3/2}\rho^{-1/2}\Theta^{3/2}$. Since on a solid or liquid planet the typical interparticle separation will be given by the Bohr radius, $e^{-2}m_e^{-1}$, the ensuing mass density will be a few times that of water, with order of magnitude $\rho \approx e^6\,m_e^3\,m_p$, so for hydrogen binding to be marginal the planetary mass must be given roughly by $M \approx 10\,m_p^{-2}(\Theta/e^2 m_e)^{3/2}$. Now, in order for water to exist in liquid form, the temperature must be small, but not extremely small, compared with the Rydberg (electronic binding) energy, and therefore given by an expression of the form:

$$\Theta_\oplus \approx \tfrac{1}{2}\,\varepsilon\,e^4\,m_e, \tag{8}$$

where ε is a numerical factor that was taken by Press to be about 3×10^{-3}. On the basis of these considerations, Press obtained for the radius of an earth like planet an expression of the form:

$$R_\oplus \approx 2\,\varepsilon^{1/2}\,e^{-1}\,m_e^{-1}\,m_p^{-1}. \tag{9}$$

This corresponds to a mass given by an expression of the form:

$$M_\oplus \approx 8\,\varepsilon^{3/2}\,e^3\,m_p^{-2} \tag{10}$$

which is smaller that the value $M_\odot \approx m_p^{-2}$ that characterises a typical main sequence star such as the Sun by a factor of order $\varepsilon^{3/2}\,e^3$.

To make the link with local biology what one needs is not the global quantities given by the preceding Press formulae, but the value of the local Galilean acceleration field given (according to the gravitation law discovered by Hooke but subsequently named after Newton) by $g \approx M_\oplus/R_\oplus^2$. By a convenient cancellation (that does not seem to have been previously noticed) it turns out that the result depends on the mass only of the electron, not the proton, taking the remarkably simple form:

$$g = 2\,\varepsilon^{1/2}\,e^5\,m_e^2, \tag{11}$$

which is only weakly dependent on the small arithmetical factor ε, but strongly dependent on the electron charge e.

7. THE WEAKNESS OF ANIMAL FLESH

Having thus discovered what governs the value of our local gravitational field, we are now faced with the less simple question of how the resulting value of g affects biological organisms of the kind to which we belong. This is an issue that I think deserves detailed biophysical investigation. Deviating at this point from the approach followed by Press [26] – and also from a related approach recently developed by Page [27] – my own suggestion is that one should think in terms of a basic biological – or to be more specific, zoological – characteristic velocity $\tilde{v}$, given, as a small fraction of the speed of ordinary sound at the relevant temperature, by a relation of the form:

$$\tilde{m}\tilde{v}^2 \approx \Theta_\oplus , \tag{12}$$

in which $\tilde{m}$ is a mass scale characterising relevant large biochemical molecules such as proteins. This means that it will be expressible by a relation of the form:

$$\tilde{m} \approx \tilde{\varepsilon}^{-1} m_{\mathrm{p}}, \tag{13}$$

in which (like the Press coefficient ε introduced above) the quantity $\tilde{\varepsilon}$ is a small arithmetical factor that – for $\tilde{m}$ to be the mass of a typical protein molecule – should have order of magnitude $\tilde{\varepsilon} \approx 0.3 \times 10^{-4}$.

This quantity $\tilde{\varepsilon}$ has the same status as that of the Press coefficient ε, in that the smallness of these quantities is just an arithmetical (in principle calculable) measure of the complexity of the (molecular, not just atomic) systems involved, and – contrary to a notion that was suggested, but justifiably criticised by Peierls, in a related discussion [28] – it has nothing to do with the smallness of the fine structure constant e^2 (whose role is significant only when heavy metals are involved) nor of the ratio $m_{\mathrm{e}}/m_{\mathrm{p}}$ (which is too small to matter much except for the lightest elements, namely pure hydrogen and helium, at temperatures far too low for ordinary life).

The foregoing estimate for $\tilde{\varepsilon}$ is actually interpretable as meaning that the corresponding zoological characteristic velocity,

$$\tilde{v} \approx e^2 (\tilde{\varepsilon}\,\varepsilon\,m_e\,/\,m_p)^{1/2},$$ (14)

will in our case be about 3 percent of the speed of ordinary sound. This relatively slow speed is mainly attributable to the very small value of the foregoing estimate for $\tilde{\varepsilon}$, which has nothing to do with the values of physically adjustable coupling constants, but is an ineluctable concomitant of the flabbiness of flexible flesh, or even cartilage, as contrasted with the rigidity of woody celluloid matter in plants, for which the corresponding botanical characteristic velocity would be considerably higher, though still subsonic. (To emulate the strength of vegetable matter, animal bodies do of course incorporate rigid bone structure, but the extent to which that is feasible is limited by the ensuing sacrifice of flexibility and mobility. I therefore differ from Press [26] in my opinion that the properties of bone itself are of secondary importance, and that the essential restrictions on animal size are attributable to the finite strength of the flexible tissues that hold the bones in place).

Assuming that such a velocity $\tilde{v}$ (of the order of 10 m/sec in our own case) characterises the relevant energies, pressures, and tensions (as involved for example in the pumping of blood) in an animal body, it will provides a rough upper limit,

$$2g\ell \lesssim \tilde{v}^2,$$ (15)

on the supportable value of the gravitational energy per unit mass associated with a height difference ℓ between different parts of the body of the organism. (When applied to solid crystalline matter, for which the relevant velocity will be of the order of that of sound, analogous reasoning correctly indicates that the maximum possible height [29] for a terrestrial mountain will be of the order of 10^4 metres).

It is instructive to see how the limit (14) can be derived in a manner similar to that used by Press [26], who considered the total energy, $\tilde{E}$ say, needed to break the bonds with energy $\Theta_\oplus$ binding the molecules on a 2-dimensional shear-disruption surface with area ℓ^2. In terms of the relevant molecular dimension $\tilde{a}$ say (which for the large protein molecules considered here will be some tens of times larger than the Bohr radius $e^{-2}m_e^{-1}$) the number of molecules in the surface layer will be of order $(\ell/\tilde{a})^3$ which gives $\tilde{E} \approx \Theta_\oplus (\ell/\tilde{a})^2$. This disruption energy has to be supplied by the action of gravity on the mass, $\tilde{M}$ say, in the corresponding volume of order ℓ^3, which will be given by $\tilde{M} \approx \tilde{m}(\ell/\tilde{a})^3$. The viability condition proposed by Press was that the energy liberated in an animal's fall through its own height ℓ should be insufficient to provide the disruption energy, which gives a limit of the form $g\,\tilde{M}\,\ell \lesssim \tilde{E}$.

My own opinion is however that animals can learn how to take care to avoid such dynamical falls, and that what really matters for viability is the condition for static support against gravity, which will hold so long as the disruption energy cannot be provided by a displacement comparable with half the molecular separation distance $\tilde{a}$, which means that instead of the preceding Press type inequality one gets a limit of the form $g\,\tilde{M}\,\tilde{a}/2 \lesssim \tilde{E}$, with the factor ℓ replaced by $\tilde{a}/2$. This replacement does not of course eliminate the dependence on ℓ, which is implicitly involved through both $\tilde{E}$ and $\tilde{M}$. Ultimately, it is the dependence on $\tilde{a}$ that cancels out, leaving a static support condition of the simple form (14).

This simple result contrasts with what would be obtained from a dynamical survivability condition of the kind proposed by Press, which reduces to the not quite so simple form:

$$g\,\ell^2 < \tilde{a}\,\tilde{v}^2.$$

For the actual application [26] of this latter formula to hard bone – instead of the fleshy tissue considered here – the molecular radius $\tilde{a}$ has to be replaced by the ordinary Bohr radius $a_0 = e^{-2}m_e^{-1}$ and the very low value of the zoological velocity $\tilde{v}$ used here has to be replaced by the much larger velocity value that is obtainable from (12) or (13) simply by setting $\tilde{\varepsilon}$ to unity.

8. CHARACTERISTIC TIMESCALE OF HUMAN PERCEPTION

On the basis of the static support condition (14), the preceding considerations imply that (whereas a land plant or a sea animal may be able to be larger) a land animal will be able to have at most a maximum size $\tilde{\ell}$ and a corresponding biological clock timescale $\tilde{\tau}$ given by:

$$\tilde{\tau} \approx \frac{\tilde{\ell}}{\tilde{v}} \approx \frac{\tilde{v}}{2g}. \tag{16}$$

Using the preceding estimates (10) and (13) for g and $\tilde{\upsilon}$ one obtains the formula:

$$\tilde{\tau} \approx 1/(4e^3 \, m_e^{3/2} \, \tilde{m}^{1/2}), \qquad\qquad (17)$$

which is independent of the previously introduced Press coefficient ε. It does however have a weak dependence on the newly introduced coefficient $\tilde{\varepsilon}$ when expressed in terms of the mean coupling constant $\langle m \rangle^2 \approx 10^{-40}$ defined by (2), taking the form:

$$\tilde{\tau} \approx \frac{(\tilde{\varepsilon} \, m_p / m_e)^{1/2}}{4e^3 \langle m \rangle^2} \qquad\qquad (18)$$

Since the factor $1/4e^3$ is only of the order of a hundred, and the dimensionless combination $\tilde{\varepsilon} \, m_p / m_e$ can be expected to be rather smaller than unity, it is the factor $1/\langle m \rangle^2$ that is overwhelmingly dominant, so the expectation of an essentially gravitational explanation is confirmed.

9. STRONG ANTHROPIC REASONING

The idea of what I called the strong anthropic principle [6] was to extend the arena of application of the weak anthropic principle to scenarios in which the relevant (anthropically weighted) a priori probability measure is not limited to the part of the universe about which we have direct observational knowledge, but extended to other hypothetically existing parts of what may be termed a multiverse [7], in which fundamental parameters, such as the fine structure constant e^2 and the gravitational coupling constant $\langle m \rangle^2$, might have values different from those (respectively 1/137 and 10^{-40}) with which we are familiar. In particular, I suggested that given the value of the former (electromagnetic) coupling constant, the weakness of the latter (gravitational) coupling constant might be explicable in such a framework as due to a selection effect – giving it the maximum value, proportional to a very high (the twelfth) power of the fine structure constant – on the basis of the requirement of stellar convection as a prerequisite for the necessary planetary formation.

Starting with the application to the marginal nuclear binding condition (1), arguments of this strong anthropic kind have since been put forward [30-32] to account for relations involving other parameters, such as those controlling weak interactions. However – in the absence of plausible mechanisms for the cosmological parameter variations that had to be invoked – such explanations did not become fashionable until the situation was revolutionised [9, 10] on the theoretical side by the rise of modern superstring theories and the attempts to unify them in M-theory, and on the observational side by the discovery (which surprised everyone, including superstring theorists) that the expansion of the universe is not undergoing gravitational deceleration but actually accelerating.

The logic leading to the formula (17) for the zoological clock timescale $\tilde{\tau}$, and to the corresponding space dimension,

$$\tilde{\ell} \approx \varepsilon^{1/2}\, \tilde{\varepsilon} / (4 e m_e m_p) \tag{19}$$

is based on the assumption that, natural selection will tend to maximise the latter (within the limits imposed by the ambient Galilean gravitational field g) in order to obtain as large as possible a value for the corresponding corporal particle number, as given by $\tilde{N} \approx \tilde{n}\,\tilde{\ell}^3$ in terms of the particle number density $\tilde{n} \approx (e^2 m_e / 2)^3$ of water. The value:

$$\tilde{N} \approx (\varepsilon^{1/2}\, \tilde{\varepsilon} e / 8 m_p)^3, \tag{20}$$

thus obtained for the body – of which the brain, in the human case, is a significant fraction – will of course be a wide overestimate of what, in view of the haphazard nature of natural selection, is actually likely to be achieved in practice, but the limit given by (18) has occasionally been approached in a few gigantic cases such as that of the brontosaurus. As a fraction of the number this is expressible as the relation:

$$\tilde{N} / N_\oplus \approx (\tilde{\varepsilon} / 16)^3 \approx 10^{-17}, \tag{21}$$

in which the right hand side is a purely arithmetical quantity whose extremely small value does not depend on any empirical parameter (suchas the proton mass m_p with which it is numerically comparable) but is just a consequence of the complexity of biochemical processes. Its cube root – of the order of a millionth –

characterises the maximum size ratio, $\tilde{\ell} / R_{\oplus}$, for a land animal with the same kind of biochemistry as ours, not just on any inhabitable planets within our own or nearby galaxies, but even in other parts of the multiverse(where quantities such as e and m_{p} need not have the values that are familiar). More particularly and memorably, on any such planet, the maximum land animal size $\tilde{\ell}$ will be of the order of a thousandth of the atmospheric thickness (and the maximum mountain height) which – as remarked above in the discussion of (14) – will itself be of the order of a thousandth of the planetary radius $R_{\oplus}$ given by (8).

In conclusion, assuming that mental processing benefits from maximisation of the number of particles in the brain and therefore of its host body, it can be seen from (19) that the strong anthropic principle will favour regions of the multiverse in which the ratio e/m_{p} is very large, that is to say where the ratio of gravitational to electric coupling is very small.

ACKNOWLEDGEMENTS

I wish to thank Don Page for many discussions that have helped me to clarify the various issues dealt with here.

CONFLICT OF INTEREST

The author(s) confirm that this chapter content has no conflict of interest.

REFERENCES

[1]   Carter, B., "Anthropic principle in cosmology", in *Current issues in Cosmology*, ed. Pecker, J.C., Narlikar, J., (Cambridge U.P. 2006) 173-179; arXiv: gr-qc/0606117.
[2]   Carter, B., "Anthropic interpretation of quantum theory", *Int. J. Theor. Phys.* 43 (2004) 721-730; arXiv: hep-th/0403008.
[3]   Carter, B., "Micro-anthropic principle for quantum theory", in *Universe or Multiverse?*, ed. Carr, B.J., (Cambridge U.P. 2007) 285-319; arXiv: quantum-ph/0503113.
[4]   Dyson, F.J., "Time without end: physics and biology in an open system", *Rev. Mod. Phys.* 51 (1979) 447-460.
[5]   Page, D., "Sensible quantum mechanics: are probabilities only in the mind?", *Int. J. Mod. Phys.* D5 (1996) 583-596; arXiv: gr-qc/9507024.
[6]   Carter, B., "Large Number Coincidences and the Anthropic Principle in Cos-mology", in *Confrontations of Cosmological Theories with Observational Data* (I.A.U. Symposium 63) ed. Longair, M. (Reidel, Dordrecht, 1974) 291-298.

[7]   Davies, P.C.W., "Multiverse cosmological models", *Mod. Phys. Lett.* A19 (2004) 727-744; arXiv: astro-ph/0403047.

[8]   Carter, B., "Classical Anthropic Everett Model: Indeterminacy in a Preordained Multiverse", in *Consciousness and the Universe: Quantum Physics, Evolution, Brain and Mind*, ed. Penrose, R., Hameroff, S., Kak, S., (Cosmology Science Publishers, Cambridge, Mass., 2011) 1077-1086; arXiv: 1203.0952.

[9]   Kallosh, R., Linde, A., "M theory, cosmological constant, and anthropic principle", *Phys. Rev.* D67 (2003) 023510; arXiv: hep-th/0208157.

[10]   Kallosh, R., "M/string theory and anthropic reasoning", in *Universe or Multiverse?*, ed. Carr, B.J., (Cambridge U.P. 2007) 191-210.

[11]   Susskind, M., "The anthropic landscape of string theory", in *Universe or Multiverse?*, ed. Carr, B.J., (Cambridge U.P. 2007) 247-266; archiv: hep-th/0302219

[12]   Carter, B., "The significance of numerical coincidences in nature", *preprint* (DAMTP Cambridge, 1967), arXiv: 0710.3543 [hep-th]

[13]   Dyson, F.J., "Energy in the universe", *Sci. Am.* 225 (1971) 50 - 59.

[14]   Pochet, T., Pearson, J.M., Beaudet, G., Reeves, H., "The binding of light nuclei, and the anthropic principle", *Astron. Astroph.* 243 (1991) 1-4.

[15]   Bondi, H., *Cosmology* (Cambridge U.P., 1960).

[16]   Carter, B., "Hominid evolution: genetics *vs.* memetics", *Int. J. Astrobiology* 11 (2012) 3-13; arXiv:1011.3393.

[17]   Dicke, R.H., "Dirac's cosmology and Mach's principle", *Nature* 192 (1961) 440-441.

[18]   Carter, B., "Mechanics and equilibrium geometry of black holes, membranes, and strings", in *Black Hole Physics*, ed. de Sabbata, V., Zhang, Z., (Kluwer, Dordrecht, 1992) 283-357; arXiv: hep-th/0411259.

[19]   Hartle, J.B., Srednicki, M., "Are we typical", *Phys. Rev.* D75 (2007) 123523; arXiv: hep-th/0704.2630.

[20]   Dyson, F.J. "Reality bites", *Nature* 380 (1996) 296.

[21]   Leslie, J., "Time and the anthropic principle", *Mind* 101 (1992), 521-540.

[22]   Carter, B., "The anthropic principle and its implications for biological evolution", *Phil. Trans. Roy. Soc.* A310 (1983) 347-363.

[23]   Lineweaver, C.H., Davis, T.M. "Does the rapid appeareance of life on earth suggest that life is common in the universe?", *Astrobiology* 2 (2002) 293-304; arXiv: astro-ph/0205014.

[24]   Carter, B. (2008) "Five or six step scenario for evolution?", *Int. J. Astrobiology* 7, 177-182. [arXiv:0711.1985]

[25]   Watson, A.J. (2008) "Implications of an anthropic model of evolution for emergence of complex life and intelligence", *Astrobiology* 8, 175-185.

[26]   Press, W.H., "Man's size in terms of fundamental constants", *Am. J. Phys.* 48 (1980) 597-598.

[27]   Page, D.N. "The Height of a Giraffe", arXiv: 0708.0573.

[28]   Press, W.H., Lightman, A.P., "Dependence of macrophysical phenomena on the values of the fundamental constants", *Phil. Trans. R. Soc. Lond.* A310 (1983) 323-336.

[29]   Weisskopf, V.F., "Search for simplicity: mountains waterwaves and leaky ceilings", *Am. J. Phys.* 54 (1986) 110-111.

[30]   Carr, B.J., Rees, M.J., "The anthropic principle and the structure of the physical world", *Nature* 278 (1979) 605-612.

[31] Hogan, C.J., "Why the universe is just so", *Rev. Mod. Phys.* 72 (2000) 1149-1161; arXiv: astro-ph/9009295.

[32] Page, D.N., "Anthropic Estimates of the Charge and Mass of the Proton", arXiv: hep-th/0302051.

Send Orders of Reprints at reprints@benthamscince.net

CHAPTER 5

The Role of Biological Time in Microbial Self-Organization and Experience of Environmental Alterations

Gernot Falkner* and Renate Falkner

Neurosignaling Unit, Cell Biology Department, University of Salzburg, Salzburg, Austria

> For us believing physicists, the distinction between past, present and future is an illusion, even if a stubborn one.
>
> *Albert Einstein*

> Zeit ist, wenigstens potentiell, die höchste, nutzbarste Gabe, in ihrem Wesen verwandt, ja identisch mit allem Schöpferischen und Tätigen, aller Regsamkeit, allem Wollen und Streben, aller Vervollkommnung, allem Fortschritt zum Höheren und Besseren.
>
> *Thomas Mann, "Lob der Vergänglichkeit"*

Abstract: A notion of biological time can be derived from distinct physiological 'units of becoming', by which an organism recreates itself in individual acts of experience. Using this starting point we present characteristic features of biological time in lower organisms. In these organisms the irreversibility of every act of experience can be attributed to a sequence of physiological processes of specific duration, termed "adaptive events", in which energy converting subsystems of a cell interact with the changing environment. In this process, the subsystems pass, *via an adaptive operation mode*, from one *adapted state* to the next. In adaptive operation modes alterations of the environment are 'interpreted' in the light of former experiences. Thereby, the subsystems are reconstructed according to these interpretations such that the resulting adapted state potentially allows optimal performance of the organism in the future. In this regard adaptive events contain a temporal vector character in that they connect former with future events and establish the irreversible historicity of the life process, in which information pertaining to the self-preservation is transferred from one adaptive event to the next: the latter "inherits" the results of former interpretations. By appropriating it selectively, it is entering into a future in which its own interpretation is

*Address correspondence to Gernot Falkner: Neurosignaling Unit, Cell Biology Department, University of Salzburg, Salzburg, Austria; Tel: + 43 (0) 662 8044 5550; Fax: + 43 (0) 662 8044 619; E-mail: Gernot.Falkner@sbg.ac.at

passed on to the following adaptive event. This system-theoretical line of reasoning is elaborated in detail using an example from aquatic ecology. In a generalization for higher organisms the temporal vector character of adaptive events is related to basic propositions of Whitehead's process philosophy and to the time concept of Augustinus.

Keywords: Adaptive behavior, autopoiesis, biological time, cyanobacteria, duration, energy conversion, experience, organismic intentions, self-referential systems, self-constitution.

1. THE ANALOGIES BETWEEN EXPERIENCE AND PHYSIOLOGICAL ADAPTATION AS A KEY TO A CONCEPT OF BIOLOGICAL TIME

1.1. The Deficiencies of Classical Physics for an Understanding of Biological Self-Organization

Classical physics operates with a concept of time that was originally given by Newton in his famous *Scholium*. Accordingly time is supposed to "flow equably without regard to anything external" (quoted from [1], p. 70), consisting of a linear succession of self-contained instants. This concept created for comparing and measuring the motion of independently existing things in a three-dimensional space was a precondition for selecting and investigating reproducible phenomena and allowed establishing the field of experimental classical physics, consisting of a catalogue of protocols by which *identical* experiences are made by an observer. Naturally this approach ignores all phenomena, in which historicity and the irreproducible creativity of a *natura naturans* plays an important role in organismic self-organization. The notion of time of classical physics, forming a continuum that is separable into infinitesimal quantities which do not penetrate each other and from which all temporal qualities such as past, present and future have been abstracted, was in physics a precondition for the development of laws of nature using non-temporal natural constants.

In this regard physical time shares a characteristic analogy with a notion of substance based upon materialism. This notion bears the essential features of the Cartesian *res extensa* in that reality is considered as a togetherness of independently existing entities, which are indifferent to each other and therefore have no feelings for each other. An application of this concept to biology ultimately tries to explain the functioning of living systems by the interaction of distinct cellular constituents. Consistent with this idea is a dismembering research strategy by which these components are first isolated and then studied under conditions, under which an internal relatedness to the living whole of an organism no longer exists. Setting out from the mathematical construction of physical time and the idea of "independently existing substances with simple location" ([2], p. 61), mechanistic physiology is obliged to postulate that living systems have states, characterized by the spatial arrangement of molecules, and that "upon these states some kind of dynamical laws, or equations of motion are superimposed" [3]. Accordingly, each instant in physical time corresponds to a locally defined state that represents a certain configuration of matter and biological processes are supposed to occur on trajectories between theses states. Such a trajectory-based view obscures the nature of biological processes, for the following reason: firstly, the imagination that organisms assume definite structures, each of them representing a state of nature at an instant of time, is an illusion; it results from the properties of our highly developed sensory apparatus, attributing structures to slower changes of a living system and processes to faster ones. Secondly, a consideration of forces acting on localized states of living subsystems ignores the fact that during a transition from one experimentally objectified 'state' to another the subsystem in question is frequently totally reconstructed; it therefore does not exist in this period as an entity that is meaningful for the organism. In such a case a transition can hardly be attributed to definable trajectories that describe the directionality of the reconstruction process (see below). Since the individual termination of this process requires certain duration for its completion, it cannot be dissected into independently existing parts along an axis of time. A reduction of the behaviour of an organism to the interaction of its molecular components does not solve this problem. A disintegration of an organism to an ensemble of chemical compounds does not allow for a sharp distinction between an organism and its environment, since molecules do not carry a label, indicating to which region they belong. Therefore the reductionist view cannot tell which state of a subsystem is useful for future activities of an organism in a given environment, rendering impossible a characterization of self-organization processes that occur in response to an environmental alteration.

1.2. The Physiological Basis of Biological Time

In contrast to physical time that can be measured objectively by a clock, biological time is experienced subjectively as duration, *i.e.,* as a present whole that contains within itself a past and a future. Naturally such a notion of biological time is not an issue for contemporary mechanistic and objectivistic physiology, which does not account for the psychophysical aspect of experience. A suitable metaphysical basis for biological time can be derived from process philosophy in which the experience of an organism is considered as a self-creative process. This conception of experience, elaborated in more detail in Whitehead's philosophy of organism [1], has recently become the basis of a system/environment theory. Of particular interest for a treatment of biological time is Luhmann's adoption of the concept of *autopoiesis* [4] for a theory of the relation between *self-referential systems* and their environment [5]. It must be noted, however, that the idea of interdependence between experience and self-constitution has also been evoked by other philosophers of the 20th century. In a historical review the American philosopher John Dewey made a major contribution to this problem. He suggested seeking in human experience for analogies to the process of physiological adaptation and that these analogies be used for a better understanding of the psychophysical dimension of life activities [6].

In his book 'Experience and Nature' Dewey approached the temporal aspect of the relation between experience and physiological adaptation by elaborating the difference between organisms and non-living things in that "the activities of the former are characterized by needs, by efforts which are active demands to satisfy needs, and by satisfactions" ([6], p. 252). The physiological correlate of that goal-oriented chronology of need, effort and satisfaction, representing the most elementary time-unit for an act of experience, was found by Dewey in an irreversible transition from one stationary state of the organism to the next. Accordingly, 'need' appears when an energetically favourable state is disturbed by an external influence, producing a "tensional distribution of energies such that the body is in a condition of uneasy or unstable equilibrium". The response to that "unstable equilibrium", manifesting itself as 'demands' and 'efforts' of the organism, leads to modifications of the environment in ways which react upon the organismic body, such that its characteristic pattern is restored by a new but different stationary state. The resulting "recovery of equilibrium pattern, consequent upon changes of environment due to interactions with the active demand of the organism" is then experienced by the organism as 'satisfaction' ([6], p. 253).

Dewey's analysis can be given a physiological meaning when the described 'equilibrium pattern' is interpreted as an organism-specific state of minimal entropy production, attained under the prevailing environmental conditions. Since during the described chronology of need, demand and satisfaction the environment of the organism is altered, every restored stationary state represents an entirely new organismic structure. The modifications of the environment, caused by the demands of the organism, occur in a self-referential manner in which the organism determines, in what respect the interaction with its environment "has significance for itself" ([1], p. 25). The reconstruction of the organism during such a self-referential cycle is accomplished by a sequence of intracellular *adaptive events* in which energy converting subsystems, as described in the following section, conform to each other in order to attain a state of least energy dissipation. An adaptive reconstruction during a new act of experience is initiated, when an alteration of the external milieu disturbs an energetically favourable operation of at least one of the energy converting subsystems of the organism. In response to this disturbance the subsystem in question then adapts to the altered external conditions by adopting new properties that lead to altered substrate and product concentrations (see below). But then the other energy converting subsystems of the cell are no longer conformed to the altered product concentration and have to readapt, presupposing that the other subsystems also develop towards states of least energy dissipation. This may again affect the initially attained adapted state and lead here to a new adaptive response, and so on. As long as the subsystems are not sufficiently conformed to each other, the total system, *comprising of the cellular system and its environment*, is remote from a balanced condition that here functions as an attractor for the overall adaptive dynamics. The whole process comes to an end, when all energy converting subsystems have been conformed to each other, which requires a defined duration. By recurrent repetitions of such elementary time segments, each of them corresponding to an act of experience, an organismic system constantly reproduces itself, concomitantly with a degradation of previously produced constituents. The result of this activity is a self-sustaining *unitas multiplex* in which each adaptive event reflects the difference between system and environment, albeit from a different perspective. In this way a system distinguishes itself from its environment by self-differentiation (a physiological example is given below). Self-reference of these operations is responsible for all kinds of manifestations of living systems, such as maintenance of a certain form or a distinct behaviour.

The proposed model explains the analogy between experience and physiological adaptation: on the one hand, every experience proceeds in the context of a prehistory of experiences, being determined by events that are learned by the organism, because they have played a certain role in its development. This corresponds to the observation that every adaptive response is influenced by the outcome of preceding adaptations, so that in this respect an organismic memory is revealed in adaptive processes. On the other hand, experience and physiological adaptation are endowed with an anticipatory orientation towards the future: experiences are determined by intentions; physiological adaptation strives for a final state, in which the living system has been conformed to *its* environment. For this reason Whitehead postulated the existence of a 'subjective aim', as a basis for final causation ([1], p. 87). Using a generalized notion of experience, from which all traits characteristic for higher animals only have been removed, allows relating experience to adaptive processes even in lower organisms, where a physiological characterization can be performed easier than in more complex organismic systems. The subsequent treatment is based predominantly on observations of microorganisms, but can be generalized for higher organisms.

A notion of biological time can only be based on analogies between experience and physiological adaptation, when the question is answered how particular acts of experience consist of distinct intracellular adaptive events and how these events may be differentiated from the constant flux of structural alterations that characterize every living being. This can be achieved by distinguishing between two distinct manifestations of an adaptive event, namely *adaptive operation modes* and *adapted states*. The adaptive operation mode refers to the above described transition of an energy converting subsystem from one stationary state to the next, caused by an environmental influence. In this mode an organism 'subjectively interprets' (*i.e.,* in a self-referential manner, see above) a deviation from a stationary state in respect to an alteration of the pre-existing organismic structure. The result of this interpretation, guided by the present 'tensional distribution of energies', leads to a new adapted state of a cellular constituent, which can then be objectified in a new interpretation. The ontological difference between subjective interpretations, occurring in adaptive operation modes and the objective features of adapted states is a consequence of the psychophysical dimension of every act of experience. It is an essential precondition for information processing about environmental alterations by adaptive events. It is obvious that these events can only be found in a dynamic manifestation of the whole organism and not in a predetermined structural set-up, such as the genome. In the following section the metabolism of a cell and the energy flow there-in are elucidated in respect to potential information processing.

1.3. Information Processing by the Adaptive Properties of Energy Converting Subsystems

The energy flow through a metabolic system, involved in the formation and degradation of all constituents of an organism, is accomplished by a great number of energy converting subsystems. For example, energy converting subsystems are the respiratory or photosynthetic electron transport chain in mitochondria, thylakoids of bacteria, in which the electron flow is coupled to a variable degree with the biosynthesis of ATP, the universal energy source for all kinds of energy dependent biochemical processes, each of them also representing an energy converting subsystem. Furthermore, the so called 'two-component regulatory systems', monitoring external signals, participate *via* phosphorylation- and dephosphorylation-reactions in the energy flow through a metabolic system. By 'cross-regulation' among different signal transduction pathways an information processing network can be established, coordinating the biosynthesis of different proteins with the subsequent growth requirement of a cell in a great many ways. Moreover, cytoskeleton biosynthesis in growing and moving eukaryotic cells depends on ordered interplay of many energy converting subsystems, potentially involved in information processing. Last, but not least, a whole microbial cell can be considered as a single energy converting unit, when all of its intracellular energy converting subsystems are conformed to each other such that available substrates and energy are utilized with optimal efficiency under the prevailing environmental conditions.

All energy converters have in common that they consist of a driving input flow and a driven output flow, as represented in a simplified scheme in Fig. **1**. The efficiency of energy conversion depends on the degree of coupling between these two flows. For each substrate concentration [A] there is only one degree of coupling, at which energy dissipation is minimal at stationary states that are conformed to the metabolic requirement of the cell [7-9].

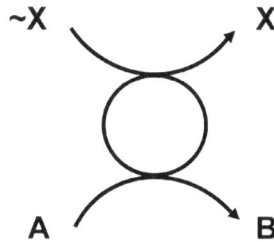

Figure 1: Model of an energy converter in which the conversion of a high energy compound ~X to a low energy compound X drives an energy consuming reaction A→ B.

When in such a state of optimal efficiency the concentration of [A] is altered, dissipation of energy increases and a new energetically favourable adapted state can only be attained, when the degree of coupling is readjusted. For this purpose the energy converter is reconstructed in an adaptive operation mode, in which usually not only the energetic, but also the kinetic properties of the energy converters are modified. The reconstruction that occurs in the light of antecedent experiences (an example is given in the following section) influences the product concentration [B], which affects the cellular metabolism. To what extent this is the case, depends on the new kinetic properties of the energy converter. In consequence, other energy converting subsystems of the cell have to adjust their energetic and kinetic properties to the new situation, because otherwise the cell would not operate in an energetic favourable manner. The resulting interplay constitutes a nexus (in Whitehead's terminology, [10], p. 201) that bears features of an intracellular communication in which each subsystem alternatively interprets the objectified adapted states of the other in regard to a potentially useful function [11]. For this reason an intracellular communication strives for a scenario in which all subsystems are conformed to each other, such that the overall process occurs with minimum entropy production. In this way communication is the basis for a permanently occurring autopoietic reconstruction of the cell [4, 5].

In adapted states an ensemble of coupled stationary flows remains approximately constant in their relation to each other for a certain period of time. The metabolic manifestation of these flows can then potentially be perceived by an external observer as (part of) a time-invariant organismic structure (presupposing that the structure is not affected by the observation). If, however, a new degree of coupling is established in an adaptive operation mode, a previously conformed ensemble of components of an energy converter is replaced and ceases to exist as part of the coupled flows of a cell. The transition from one adapted state to the next requires a defined period of time that has to be fulfilled, before another energy converting subsystem can respond to the new adapted state. The dependence of an adaptive operation mode on former experiences (see below) renders impossible a prediction of the duration and direction of adaptive events and establishes the historical dimension of biological processes. Although an adaptive operation mode has a defined temporal extension, it cannot be divided in independently existing subunits. "If you abolish any part, then that whole is abolished" ([1], p. 288). Naturally this is a consequence of the fact that there are no half-adapted states and no half-made experiences.

In this regard biological time refers to temporal units of becoming that may be considered as 'drops of experience' ([1], p. 18) in which the transition from one adapted state to the next *via* an adaptive operation mode can be analyzed into three phases: first, perception of an environmental alteration, performed by a subsystem in the previously attained adapted state; second, subjective interpretation of this environmental alteration in the light of antecedent experiences; third, reconstruction of the subsystem according to this interpretation, guided by an "ideal of itself", as defined below. Naturally the three phases are interdependent, because each of these phases is part of one adaptive event that requires the whole process.

The energetic constraint to operate with optimal efficiency has an important ecological implication. During fluctuations of the external substrate concentrations, as they normally occur under natural conditions, available energy is utilized with optimal efficiency only when the cells are able to conform to the average concentration level of a given pattern of substrate fluctuations. This presupposes that cells have the capacity to process information about the prevailing pattern and to 'decide' on possible alterations of that pattern in regard to a necessary reconstruction. To what extent differences in a given pattern are recognized by the biological system depends on the reaction time that determines how the fluctuating substrate concentration is dissected in a sequence of adaptive events. In this regard the reaction time is the basis for a 'memory' of previously performed pattern interpretations along a historic succession of adapted states and adaptive operation modes.

In the following section we give an example of pattern recognition by a bacterial population, using an example from aquatic ecology. In this example information processing and memory phenomena are studied by observing the effect of antecedent on subsequent adaptive events. The empirical investigation of this phenomenon departs from an objectivistic research strategy that is inspired by the ideals of classical experimental physics, consisting of a catalogue of protocols by which identical experiences are made by an observer.

2. DISCUSSION OF THE FOREGOING CONSIDERATIONS USING A PARTICULAR BIOLOGICAL EXAMPLE

2.1. The Microbial Experience of Environmental Phosphate Fluctuations

Traditional plant physiology is confined to characterizations of adapted states. The transition of one adapted state to another – *via* an adaptive operation mode, as defined above – is usually not studied, presumably because this process evades an objectivistic description. An analysis of the adaptive behaviour of individual energy converting subsystems is also impaired by their entanglement with other adaptive processes. Investigation of the vectorial temporal aspects of adaptive events in respect to memory, information processing, possible decision making and self constitution under ever changing environmental conditions requires an appropriate bacterial model system in which these phenomena occur under defined experimental conditions. Due to the special life situations of algae and cyanobacteria, the physiological adaptation of cyanobacteria to alterations in phosphate supply is exceptionally suitable for this purpose. In oligotrophic lakes algae and cyanobacteria are exposed to extreme nutrient fluctuations. Occasional increases of the external phosphate concentration alternate with periods in which the external concentration is so low that uptake of phosphate ceases for energetic reasons [12, 13]. As a result of this energetic constraint, fluctuations of the external concentration are experienced by the cells as pulses. In this situation continuous growth of algae and cyanobacteria is only provided when the organisms are able to utilize efficiently short-term rises of the external phosphate concentration and to store this nutrient in form of polyphosphate granules. The proliferation of the organisms then proceeds at the expense of stored polyphosphates. However, this survival strategy can only work when the growth rate is conformed to the pattern of phosphate fluctuations that occur in the prevailing environment. This is achieved by a complex concatenation of adaptive events, in which information about adaptation to antecedent pulses regulates adaptation to subsequent pulses, such that the amount of accumulated phosphate in each pulse meets the demand for the anticipated growth rate, which – in turn – depends on the amount of stored phosphate. The experimental investigation of this phenomenon has shown that in this process a population of cyanobacteria is capable of discriminating between different patterns of phosphate pulses, even when during these pulses the same amount of phosphate has been stored.

An example of that kind of pattern recognition is given in Fig. **2**. Thereby, two identical populations of *Anabaena variabilis* were exposed initially to three pulses of 1, 3 and 6.5 µmoles of phosphate per litre, however, applied to one population in increasing and to the other population in a deceasing dose rate. The shape of the pulses, revealed in the time dependence of the decrease of the external phosphate concentration, reflected the prevailing uptake behaviour of the cell: the faster the phosphate disappeared in the external medium, the higher was the activity of the uptake system. After having experienced the three pulses in the two different supply modes, the cells in the two suspensions exhibited diverse uptake properties, as can be seen in a fourth pulse, applied 250 minutes after the beginning of the experiment: the population that had been exposed to an increasing dose rate (Fig. **2**, full lines), responded to the last pulse with a smaller deactivation of the uptake system than the reference population with the decreasing dose rate (Fig. **2**, dashed line). The dependence of the adaptive behaviour on the nature of antecedent phosphate exposures indicates that after each pulse a distinct adapted state emerged and that this state influenced the subsequent adaptation process in a characteristic manner. This results in an interrelationship between adaptive events, so that the adaptive response to an increase in the external concentration is determined by the pattern of previously experienced phosphate fluctuations. Information about the experienced pattern can then even be transferred to daughter generations [11, 14].

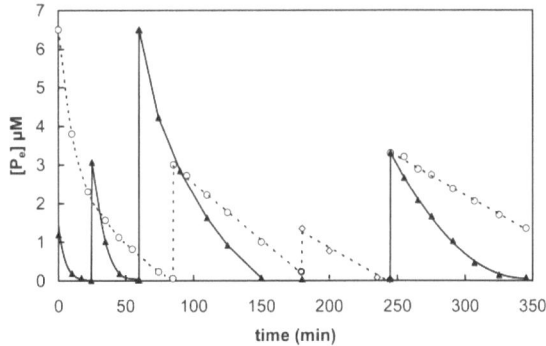

Figure 2: Time course of ^{32}P-phosphate removal from the external medium by two populations of *Anabaena variabilis*. The two different patterns of the first three pulses initiate a different adaptive behaviour during the final pulse.

The interdependent sequence of adaptive events has an important temporal aspect that is reflected in the interplay of the two energy converting subsystems, involved in the conversion of external phosphate into polyphosphates. This process proceeds in three steps (Fig. **3**):

1. Transport of external phosphate P_e from the external medium into the cell, catalyzed by a "carrier protein" (reaction 1 in Fig. **3**). At low external concentrations an energy source is needed for the translocation against the existing electrochemical gradient at the cell membrane. In the cyanobacterium *Anacystis nidulans* energy is provided by an ATPase [15], which can be coupled to the phosphate carrier to variable degrees (reaction 2 in Fig. **3**).

2. Conversion of the incorporated internal phosphate P_i to ATP *via* photophosphorylation (in the light, reaction 3 in Fig. **3**) or oxidative phosphorylation (in the dark). Also this reaction is energy dependent and potentially coupled to a variable degree to the driving process, which is here the flow of protons from the thylakoid space into the cytoplasmic space. Furthermore, the H^+/ATP stoichiometry is variable; the higher the stoichiometry, the lower the stationary phosphate concentration in the cell [15].

3.   Formation of polyphosphates from ATP. This step, in which the terminal phosphate group from ATP is transferred to polyphosphates, requires no energy (reaction 4 in Fig. **3**).

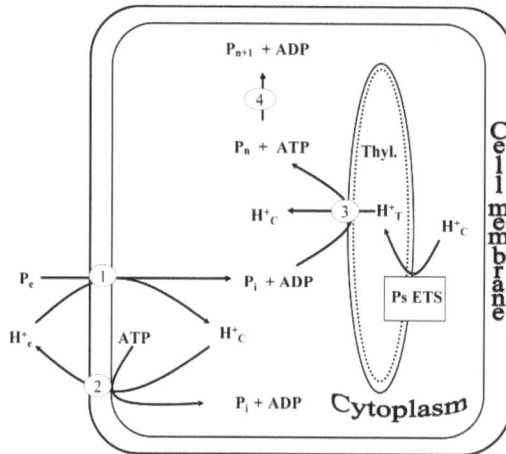

Figure 3: Schematic presentation of phosphate utilization by cyanobacteria. Reactions 1, 2, 3 and 4 show the conversion of external phosphate to polyphosphates. The correct stoichiometries are not indicated in the figure. P_e: external phosphate; P_i: internal phosphate; P_n, P_{n+1}: polyphosphates; H^+_C and H^+_T are the proton concentration in the cytoplasmic and thylakoid space; Ps ETS: photosynthetic electron transport system. For further details, see text.

Since the ATPase, associated with the phosphate carrier, receives its substrate from the ATP-synthase at the thylakoid membrane, the transport process is indirectly coupled with proton flux across the thylakoid membrane. In adapted states the degree of coupling between transport process and the proton flux is conformed to the external concentration such that energy dissipation is minimal [16]. As a result of this energetic constraint the degree of coupling is increased, when the external concentration decreases. This affects the threshold value, *i.e.,* the concentration at which the available energy no longer suffices to drive the transport through the cell membrane, because the threshold value becomes lower with an increase in the degree of coupling [17, 18]. Concomitantly with a decrease of the threshold value the activity of the transport system becomes higher [13].

When the overall uptake process proceeds with least energy dissipation, the dependence of the uptake rate J_P on the logarithm of the external concentration $[P_e]$ obeys a simple linear flow-force relationship, extending from the threshold value to the external concentration, to which the uptake system has been conformed. This relationship was originally proposed by Thellier for analyzing the concentration dependence of uptake processes in plants [19]. For the present case it attains the form: $J_P = L_P \cdot (\ln([P_e] - \ln[P_e]_A)$. L_P is a conductivity coefficient that reflects the maximum velocity of the uptake system. $[P_e]_A$ is the threshold value of the corresponding adapted state. An extended range of validity of the linear dependence of the uptake flow on the logarithm of the external phosphate concentration contradicts basic principles of non equilibrium thermodynamics; this linear dependence, however, is a pre-condition for efficient energy conversion in regions far from equilibrium and can be explained by a functional integration of high affinity and low affinity transport systems [16, 20]. It is notable that the resulting proportional flow-force relationship has the same structure as Weber-Fechner's law, when the uptake rate J_P is interpreted as the response to the stimulus $[P_e]$. Apparently adaptive events aim at establishing a mechanistically operating perceptive apparatus, in which the relation between stimulus and response follows a similar function as the sensory perception in higher organisms.

2.2. The Intracellular Communication of Two Energy Converting Subsystems

To what extent alterations of the environmental phosphate concentration are experienced and, in consequence, information is transferred from one pulse to the next, is determined by the particular features of the communication between the two subsystems. Such a communication about the changing ambient concentration is initiated by an abrupt increase of this concentration, after an onset of the uptake process at the beginning of a pulse. Naturally this process is performed first with the energetic properties of the uptake system, originating from the last pulse. When in this situation the incorporation of phosphate does not proceed with minimum entropy production, the uptake system is reconstructed, aimed at attainment of an energetic favourable state for the anticipated growth rate. In a first step the degree of coupling between the activity of the phosphate carrier and the ATPase (reaction 1 and 2) is conformed to the perceived phosphate flow into the cell, such that a state of least energy dissipation is attained. This reduces the uptake rate and prolongs the transient increase of the cytoplasmic phosphate concentration, so that the ATP-synthase gains time to conform the degree of coupling between the proton flow and the ATP-synthase to the elevated cytoplasmic phosphate concentration, aimed at an efficient operation at this elevated concentration. Concomitantly and as a result of the uptake activity by the altered properties of the two subsystems, the external concentration decreases further and the transport system has to readapt again, and so forth, until the external phosphate concentration remains constant at the finally attained threshold value, where the adaptive communication of the two systems has come to an end. The two systems are then self-referentially conformed to each other and to the external concentration which, in turn, results from that conformed state. This interdependence provides stability and explains why the distinct attained adapted properties are then maintained for a certain period of time without phosphate incorporation, establishing a kind of primitive 'memory' of antecedent exposures to phosphate fluctuations [11]. When this self-referential state is disturbed by a new pulse, intracellular communication starts afresh and is now resumed with the properties originating from the antecedent pulse. This establishes a communication tradition, in which *via* transitions between adapted states and adaptive operation modes, in a sequence of pulses information is transferred in a specific manner from one adaptive event to the next. Decisive for information processing about alterations of the ambient phosphate concentrations is the time each individual subsystem takes for adapting to the manifestations of the other. In this regard the reaction time reflects the sensitivity for external concentration changes and determines to which extent information about an environmental alteration is processed by the organism.

It is informative to discuss the temporal aspect of this communication, in which processed information depends on the interpretation of the changing external concentration by two subsystems, using a 'family tree of individual occasion of experience' [21] (Fig. **4**). Each arrowhead is a symbol for an adaptive event in which the degrees of coupling between the phosphate flow and the ATP hydrolysis on the one hand and the ATP production and the proton flow on the other hand are altered from q_n and Q_n to q_{n+1} and Q_{n+1} respectively. The lines between the arrowheads represent external (dashed lines) and internal (full lines) influences to which the individual subsystems respond by adaptive reorganizations. In this family tree the temporal passage from the past to the future, depicted in Fig. **4** by arrowheads, is a function of local experience of manifestations of antecedent states (dashed and full lines) and not the result of global temporality, in which "time flows equably without regard to anything external". Since temporality is ultimately linked with local phenomena, biological time differs fundamentally from the physical abstraction of a universal cosmic time (Mittelstaedt, this volume). Each local experience, reflecting an adaptive event, is related to other events by interplay of efficient and final causes. The outcome of antecedent adaptive events represents efficient causes for the onset of the subsequent processes. Their anticipatory interpretation, however, establishing a new cellular constituent which is supposed to provide an integer performance of the whole organism, requires explanation by some sort of 'final cause' (see below). In this respect each adaptive event is characterized by a temporal transition from potentiality *via* actuality towards a past, in which it becomes a causal factor for a new adaptive event. The interplay between efficient and final causes, referring to the terminated outcome of antecedent and the interpreting activity of subsequent events, is a necessary consequence of the irreversibility of the experience, made by each subsystem in every adaptive event and gives rise to the irreversibility of the life process. Due to the vast number of possible stable tertiary structures of a protein the components of energy converting subsystems can be conformed to each other in practically infinite different ways, thereby forming functioning-dependent structures [22]. Since, in this process, every adaptive event leaves its mark on subsequent processes in a particular manner, there is no repetition in the creative advance into the future. In this regard every adaptive event is a unit of becoming of something new and every adapted state is the emergence of novelty: an attribution of a persistent structure to an organism or a species is based on arbitrary abstractions.

The irreversibility of the communication between the two subsystems, opposed to an exhaustive mechanistic description of living systems, can only be understood in historical terms, for the following reason: When adaptation is guided by interpretations that are derived from the outcome of former adaptations, these former adaptations have to be analyzed as representing interpretations of the more remote past, and so on. This has important implications for establishing a model of organismic experience based on adaptive events. Adaptive events as a basis for biological self-organization can only be simulated by a computer model, when this model contains two interdependent levels, accounting for the 'historical nature' of adaptive processes [23]. One level has to refer to the objective manifestations of transiently attained adapted states. The other level refers to an interpretation of these manifestations during an adaptation to a new environmental situation. This temporal aspect that takes account of the eigentime of adaptive events is left to the intuition of the modeller. By adjusting the times, at which the two systems respond to each other, the programmer can model the sensitivity of the system and to what extent information is transferred from one adaptive event to the next.

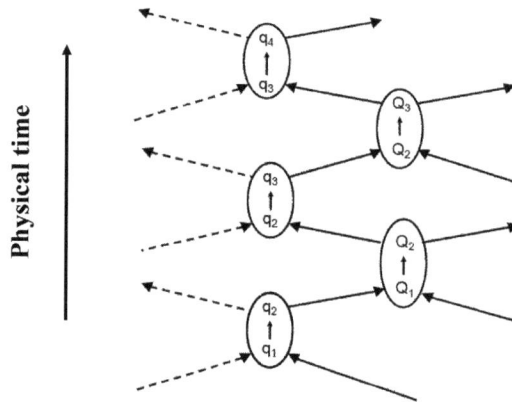

Figure 4: Family tree of interrelated adaptive events.

The impossibility of an objective characterization of adaptive events by time-invariant parameters is also a consequence of the fact that in adaptive operation modes information about environmental alterations is processed and stored by a cellular subsystem. This impairs an experimental characterization of adaptive operation modes. Since the prevailing experimental condition also constitutes information, the outcome of the investigation is influenced by the way it has been performed [24, 25]. However, due to the dependence of cellular information processing on former experiences of an organism, the response of a cellular system to experimentally imposed conditions provides an insight into the prehistory of the organism.

3. SOME FURTHER INFERENCES FROM WHITEHEAD'S PHILOSOPHY OF ORGANISM ON THE FINAL CAUSE OF ADAPTIVE PROCESSES

The system/environment theory postulates that a microorganism experiences its environment *via* a communication of a community of adaptive events, aimed at its self-constitution under ever changing external conditions. In this way intracellular communication establishes an internal relatedness among the different intracellular events and explains organismic subjectivity by the self-referential features of that relatedness [5]. Since biological time has a reference to an individual feeling of experienced durations, it is legitimate to ask whether such an internal relatedness among communicating subsystems could be – even in simple organisms – a physiological basis of some sort of 'feelings' that reflect the internal status of the cell. To answer this question, it is necessary to generalize what has been outlined above. Accordingly, an intracellular communication, initiated whenever an adapted state is lost by an environmental disturbance, may be considered as a process in which an excitation of cellular energy converting subsystems propagates in a wavy manner back and forth in all directions through the organism.

Let us now assume with Dewey that an organismic ensemble of subsystems experiences during such an excitation a state of tension and that this seemingly chaotic process (for an external observer) incites to all kinds of molecular alterations in a cell, until the tension disappears and an ordered state of intra-mundane stability emerges. According to Whitehead's philosophy of organism such a tension-free state can be considered as a defining characteristic ([10], p. 206), preserving the identity of an organism in a constant flux of metabolic alterations under ever changing environmental conditions. If so, tension would indicate that the organism has departed from its 'ideal of itself' ([1], p. 85), corresponding to a 'target value' that is anticipated in every adaptive event. Consistent with this postulate is the idea that an organism experiences itself in such a tension, which, for this reason, may be considered as an internal representation of this 'ideal of itself'. In such an "adaptive representational network" [26] the functioning of adaptive events as elements of self-constitution could then be influenced by such an intracellular tension: for example, the capacity to discriminate between environmental stimuli (as a measure for the sensitivity) would become greater when an intensification of the tension accelerates an adaptive response to these stimuli. In this regard the tension would ultimately provide connectivity between subsequent adaptive events, because a potential departure from an idealized manifestation in an antecedent event (*e.g.*, due to inappropriate interpretation of an environmental change, see above) can be corrected in a subsequent event. The resulting historical order of events is a pre-condition for the occurrence of *memory phenomena* at different levels of organization of living systems. Naturally the *self* is a logical consequence of the ontological difference between adapted states and adaptive operation modes, in which the latter determine to what extent an external influence on a given adapted state "has significance for itself", as has been termed by Whitehead ([1], p. 25). In this respect system theoretical interpretations differ from conventional objectivistic self-organization models that rest upon the time conception of classical physics and therefore lack an ontological difference between experiencing and experienced manifestations of an organismic self. These models require specific boundary conditions under which deterministic reactions occur and do not display the fact that a modeller has implemented the necessary boundary conditions. Since objectivistic models of self-organization ignore the special achievement of organisms, namely to create appropriate boundary conditions on their own under which they give themselves a new structure by deterministic reactions, they describe "self-organization without self" [27]. A process philosophical system/environment theory avoids deficiencies resulting from an objective description of biological processes by endowing the temporal vector character of adaptive events with essential features of the time concept of Augustinus. For Augustinus time is experienced as transition from expectation *via* attention to memory (Achtner, this volume). In the system theoretical concept of biological time this is converted to a 'unit of becoming', characterized by a transition from anticipation *via* a sensitivity-dependent adaptive response to stored information about that response.

The valuing decision, initiating a new adaptive operation mode, occurs at the end of an antecedent and the beginning of the subsequent adaptive event and therefore establishes a sharp line between two entities of defined time duration. Since in that moment the outcome of antecedent events is interpreted in respect to a reconstruction, in which the organism finds its way back to its identity, the dichotomizing act of 'decision making' in every interpretation is guided by something that is not subject to temporal alteration. It refers to an intentional 'feeling' of how to re-adjust all metabolic processes in every new experience, such that the resulting organismic constituents are again ideally conformed to each other. The re-adjustment occurs in a quest for an invariable appearance of the defining characteristics of the organism, providing the persistence of the phenotype of the species to which the organism belongs. "The genetic passage from phase to phase is not in physical time" ([1], p. 283). According to this idea the defining characteristic of an organism depends on the ingression of that feeling for tension-free states into the interpreting perception of antecedent occurrences. In Whitehead's process philosophy feelings have relevance as final cause determining the concrescence of subsystems to a conformed unity. In this respect adaptive events bear a characteristic analogy with Whitehead's actual entities in a way that opens a possibility to breach the gap between biology and theology: "This final cause is an inherent element in the feeling, constituting the unity of that feeling. An actual entity feels as it does feel in order to be the actual entity which it is. In this way an actual entity satisfies Spinoza's notion of substance: it is *causa sui*. The creativity is not an external agency with its own ulterior purposes. All actual entities share with God this characteristic of self-causation. For this reason every actual entity also shares with God this characteristic of transcending all other actual entities, including God. The universe is thus a creative advance into novelty. The alternative to this doctrine is a static morphological universe" ([1], p. 222). It remains to be established to what extent the evolution of more complex organisms from antecedent states of less complex organisms, reflecting a community related creation of new states of intra-mundane stabilities, can be explained along that line.

ACKNOWLEDGEMENTS

The work has been supported by the Austrian Science Fund (FWF, Project Nr. P16237-B06).

CONFLICT OF INTEREST

The author(s) confirm that this chapter content has no conflict of interest.

REFERENCES

[1]   Whitehead A. N. (1929/1978): *Process and Reality*. New York: Macmillan (corrected edition).

[2]   Whitehead A. N. (1926/1985): *Science and the Modern World*. London: Free Association Books.

[3]   Rosen R. (1985): Organisms as causal systems which are not mechanisms: an essay into the nature of complexity. In: *Theoretical Biology and Complexity*; Rosen, ed. New York: Academic Press.

[4]   Maturana H., & Varela, F. (1980): *Autopoiesis and Cognition: the Realization of the Living*. Boston: Reidel.

[5]   Luhmann N. (1995): *Social Systems*. Stanford, CA: Stanford University Press.

[6]   Dewey J. (1925): *Experience and Nature*. New York: Dover Publications, Inc.

[7]   Katchalski A., & Curran P. F. (1965): Nonequilibrium thermodynamics in Biophysics. Cambridge: Harvard University Press.

[8]   Kedem O., & Caplan S. R. (1965): Degree of coupling and its relation to efficiency of energy conversion. *Trans. Faraday Soc.* 61, 1897-1911.

[9]   Stucki J. W. (1980): The optimal efficiency and the economic degrees of coupling of oxidative phosphorylation. *Eur. J. Biochem.* 109, 269-283.

[10]   Whitehead A. N. (1933/1967): *Adventures of Ideas*. London: The Free Press.

[11]   Falkner R., Priewasser M., & Falkner G. (2006): Information processing by Cyanobacteria during adaptation to environmental phosphate fluctuations. *Plant Signaling and Behaviour*, 1, 212-220.

[12]   Hudson J. J., Taylor W.D., & Schindler D W. (2000): Phosphate concentrations in lakes. *Nature* 406, 54-56.

[13]   Falkner G., Falkner R., & Schwab A., J. (1989): Bioenergetic characterization of transient state phosphate uptake by the cyanobacterium *Anacystis nidulans*. Theoretical and experimental basis for a sensory mechanism adapting to varying environmental phosphate levels. *Arch. Microbiol.* 152, 353-361.

[14]   Falkner R., & Falkner G. (2003): Distinct adaptivity during phosphate uptake by the cyanobacterium *Anabaena variabilis* reflects information processing about preceding phosphate supply. *J. Trace Microprobe Techn.* 21, 363-375.

[15]   Wagner F., & Falkner G. (1992): Concomitant changes in phosphate uptake and photophosphorylation in the blue-green alga *Anacystis nidulans* during adaptation to phosphate deficiency. *J. Plant Physiol.* 140, 163-167.

[16]   Falkner G., Wagner F., Small J. V., & Falkner R. (1995): Influence of fluctuating phosphate supply on the regulation of phosphate uptake by the blue-green alga *Anacystis nidulans*. *J. Phycol.*, 31, 745-753.

[17]   Falkner G., Wagner F., & Falkner R. (1994): On the relation between phosphate uptake and growth of the cyanobacterium *Anacystis nidulans*. *CR Acad Sci Paris, Sciences de la vie/Life science* 317, 535-541.

[18]   Falkner G., Falkner R., & Wagner F. (1993): Adaptive phosphate uptake behaviour of the cyanobacterium *Anacystis nidulans*: analysis by a proportional flow-force relation. *CR Acad Sci Paris, Sciences de la vie/Life sciences* 316, 784-787.

[19]   Thellier M. (1970): An electrokinetic interpretation of the functioning of biological systems and its application to the study of mineral salts absorption. *Ann. Bot.* 34, 983-1009.

[20]   Wagner F., Falkner R., & Falkner G. (1995): Information about previous phosphate fluctuations is stored *via* an adaptive response of the high-affinity phosphate uptake system of the cyanobacterium *Anacystis nidulans*. *Planta*, 197, 147-155.

[21]   Hansen N. V. (2004): Spacetime and Becoming: Overcoming the contradiction between special relativity and the passage of time. In: *Physics and Whitehead*. State University of New York Press, New York.

[22]   Thellier M., Legent G., Norris V.,Baron C., & Ripoll C. (2004): Introduction to the concept of functioning-dependent structures in living cells. *CR Acad Sci Paris, Sciences de la vie/Life sciences* 327, 1017-24.

[23]   Plaetzer K., Thomas S. R., Falkner R., & Falkner G. (2005): The microbial experience of environmental phosphate fluctuations. An essay on the possibility of putting intentions into cell biochemistry. *J. Theor. Biol.* 235, 540-554.

[24]   Falkner G., Wagner F., & Falkner R. (1996): The bioenergetic coordination of a complex biological system is revealed by its adaptation to changing environmental conditions. *Acta Biotheoretica*, 44, 283-299.

[25]   Falkner G., & Falkner R. (2000): Objectivistic views in biology: an obstacle to our understanding of self-organization processes in aquatic ecosystems. *Freshwater Biology*, 44, 553-559.

[26]   Trewavas A. (2005): Green plants as intelligent organisms. *Trends in Plant Science*, 10, 413-419.

[27]   Koutroufinis S., (1996): Selbstorganisation ohne Selbst. Irrtümer gegenwärtiger evolutionärer Systemtheorien. Pharus Verlag, Berlin.

CHAPTER 6

Parts of the Brain Represent Parts of the Time: Lessons from Neurodegeneration

Hans Förstl*

Department of Psychiatry & Psychotherapy, Technical University Munich, Munich, Germany

> Nec futura sunt nec praeterita, nec proprie dicitur:
> tempora sunt tria, praeteritum, praesens et futurum,
> sed fortasse proprie diceretur:
> tempora sunt tria, prasens de praeteritis, praesens de praesentibus, praesens de futuris.
> Sunt enim haec in anima tria quaedam et alibi ea non video,
> praesens de praeteritis memoria, praesens de praesentibus contuitas, praesens de futuris expectatio.
>
> **Augustinus, Quid enim est tempus, 26**

Abstract: People get accustomed to the experience of time. An almost constant movement from the past towards the future appears psychologically inevitable and similar to gravitation, which physically and inevitably appears to pull us towards the center of the earth. This contribution describes conditions which change the perception of time; Alzheimer's disease, which eliminates the recent past; frontotemporal dementia, which diminishes curiosity, fear, and further future concerns; and delirium with a loss of a laminar stream of consciousness or time. It may be a coincidence that brain structures primarily dealing with the past rest in the back of the head, while structures considering future prospects sit in the frontal lobe, and a laminar processing of the present relies on an intact central brain structure. Even though physical time may continue as usual, its perception and validation is subject to cerebral distortions.

Keywords: Time, neurobiology, neurodegeneration, confusional state, dementia, frontotemporal lobar degeneration, dementia with lewy bodies, pick, alzheimer.

INTRODUCTION

Several theoretical positions regarding the reality and relevance of the past,

*Address correspondence to Hans Förstl: Department of Psychiatry & Psychotherapy, Technical University Munich, Munich, Germany; Tel: +49 89 4140 4200; Fax: +49 89 4140 4837; E-mail: Hans.Foerstl@lrz.tu-muenchen.de

presence and future – including their complete irrelevance and non-existence – can be summarized as follows [1]:

- Only the presence is real – but what are the limits, what is the frame of a moment?

- Presence and future are real, but not the past – this position appears biologically interesting, as our survival and success depends on a clever planning and coping with future tasks, which is basically only backed up by past experience.

- The past – but not the future – is a part of our presence and both are real and relevant. Past and presence offer the basis of different options for future developments as we experience them. But if this was true, would it imply that our experience and reality had to expand constantly with time?

- And finally: the past and the future are realms of our reality just like the presence, "determined at any moment by our temporal perspective" [1, p. 86].

Dunnett's final and preferred position, which appears immediately appealing in view of most individuals' personal perceptions of how they go through their lives, recounts Augustinus' classical concept extolled in his "quid enim est tempus" [2, 3].

My clinical perspective is based on observations in prototypical neurodegenerative diseases, which affect different parts of the human brain systematically and lead to characteristic alterations of time-related experience and behavior.

THE PAST

Forgetfulness is a typical symptom early in the course of Alzheimer dementia [4]. It first affects declarative information, events and contents which have been experienced and learned, and which can – if remembered - easily be expressed with words [5]. The necessary prerequisite for remembering is the capability to learn, *i.e.,* to obtain, understand and memorize information, and therefore an intact learning apparatus for declarative information. Bottleneck structures of this apparatus in the mediotemporal lobe, entorhinal region and hippocampus, are affected by Alzheimer type plaque and neurofibrillary changes early in the course of this degenerative process. Therefore the learning of new information is impaired and these difficulties increase with ongoing neurodegeneration. During this process the intraneuronal neurofibrillary changes are far more targeted on the structures of the declarative learning apparatus than the widespread plaque pathology. Activities of daily living are first impaired by an anterograde amnesia, which can also be verified by neuropsychological tests: from the onset of the first clinical symptoms most patients find it increasingly hard to keep track of what is going on; this leads to problems remembering events which have happened in the recent past, while remote history is usually well retained and accessible. After a longer course of illness with more intense and distributed brain changes, learning difficulties become more severe and include salient new events, remembering past events proves harder and other skills are also affected.

Secondary deficits can be understood as logical consequences of amnesia. This includes problems with temporal and spatial orientation, a process which relies on constant updates of what was intended and what has happened. This unsettling uncertainty of why what who where when has done, become, happened … may cause anxiety, agitation and aggression. Attempts at the best possible interpretation of this subjectively inexplicable experience can lead to delusional reasoning and even greater cognitive and emotional turmoil. Familiar carers are not always recognized and must find a non-confronting style of comforting communication, which is not preferably based on spoken words, as they – if understood at all – would immediately be forgotten.

In advanced stages of Alzheimer dementia, large parts of the autobiographical memory are lost, which normally provide personal continuity and the identity of the self [6]. In emotional moments with high attention, patients may become briefly aware of their lost past and personhood.

PRESENT

A combination of Alzheimer-type and Parkinson pathology may lead to a "dementia with Lewy bodies". Lewy-bodies are the pathological hallmarks of Parkinson's as plaques and tangles are the characteristics of Alzheimer's. Clinically these patients suffer from the combination of a dementia syndrome together with recurring confusional states [7]. A confusional state can be considered as a turbulence of the "stream of consciousness" [8]. More verbose attempts at understanding consciousness have become popular (*e.g.,* [9]). One aspect of this stream of consciousness, the perception of the presence, of the "moment", is short-term-memory. When testing verbal or visual material, human short-term-memory includes around 7 to 8 seconds and 7 to 8 items, - one more than a chimpanzee's. Therefore man – but not monkey – can briefly retain an 8-digit phone number, if not distracted. When man is alert and attentive to make the best use of his short-term memory capacity, he can understand a complicated sentence with several clauses or reflect perceptions and intentions of several other individuals within a larger group. Outside critical situations and psychological laboratories, moments pulsating through our consciousness may last slightly shorter [10].

Nature has provided us with a smart replay function: language. Language and its related concepts offer an opportunity to mull over exquisite phenomena and enter highly sophisticated conversations with our fellow torch bearers of humanity, quantum physicists, neurobiologists and the like. With much enthusiasm and after a good deal of highly intelligent, enlightened, logical thought-processing, we are left with the good feeling of having made a rational decision well worthy of ourselves – while the huge rest of the brain has fluently resolved all the other tasks we are not so proudly aware of. Language is one prominent attempt at making moments last. But all of these reflections take their own time and mirrors are never so well aligned to reflect one moment to another without distortion. Seemingly authentic recurrences of past events are infrequent and no sign of mental health (*e.g.,* [11]).

This stream - or rather: pumping - of consciousness with its big, "conscious" pictures and its minute minimal sensations [12] corresponds to the momentary electrochemical oscillation of the whole brain. This is the activity in the human body, which consumes most energy. It includes the feeling, filtering, amplifying, comparing and eventually selecting of information most relevant to be attended to, and eventually even to be learned. Brainstem and limbic system are conducting this research and orchestrating most of the synchronized activity in the rest of the nervous system. The result is perceived as a nearly laminar, sometimes boring, but very often entertaining orchestral piece of sufficient clarity to detect and resolve potential barriers and dangers so that we can concentrate on them to find resolutions. This ordinary flow of affairs is regularly reversed during sleep, which is most easily accepted as a pleasant change. Confusional states may initially manifest as disturbances of sleeping and dreaming, but do not usually come to a full stop after opening the eyes.

The decisive deficit in dementia with Lewy-bodies is a lack of the neurotransmitter acetyl-choline, which is synthesized in the basal nucleus of Meynert. This strategically important brainstem nucleus is attacked by Alzheimer-pathology from one, and by the Parkinson-pathology from the other side. Most of the factors necessary to cherish the moment – wakefulness, attention, filtering, learning … – are critically dependent on acetyl-choline, which can in large parts of the brain only be received from the basal nucleus. No acetyl-choline, no laminar stream of consciousness, but disturbed vigilance, nightmares, symptoms even worse, and certainly no organized storage or retrieval of relevant information. Patients will not usually remember explicitly what has happened during a confusional state. Sometimes a diffuse feeling of an unpleasant period is implicitly kept.

FUTURE

Social interactions are of great relevance for humans, who are rather unimpressive creatures if left to their own devices. A large part of the brain's computational capacity is constantly devoted to controlling and improving social skills. Sociobiologists discredit such perfectionist intentions of refining co-operation as attempts to improve the balance between our personal investment and benefit quite egotistically. This cynical caricature of human cultural achievements and ethics is so mean and so low that words fail me to construct a convincing counterargument. If this is how we are, we should stop paying so much attention to ourselves and to our kin.

Therefore let me turn to the frontal lobe, which is a wide cortical area with complicated subcortical connections. A vast fronto-dorsal cortico-subcortical circuit is important for efforts called intelligence, its fronto-orbital counterpart deals with questions addressing reason, and the medial circuit produces behaviors generally associated with (free) will. Intelligence, or reason, or will are lost, when these circuits are damaged selectively. If they collaborate in an intact brain, they achieve insight into complex social situations called "theory of mind". This theory of mind or "mentalizing" is the ability to see things from a different perspective, to understand how others view my actions, how they feel about it and will probably react to my potential actions [13]. This re-spect is important for our own and our family's survival and success. Single-mindedness must not be considered as the spearhead of human evolution. Careful planning, taking others' actual views and collective experience accumulated over time into account, means that an individual needs not make every painful experience by himself. Imagining potential consequences of my actions more frequently than not leads to inhibition, to refrain from a certain action (see the German words "Ver-Halten, An-Stand, … Hemmung"). The antagonism between drive and brake, between the so-called "pleasure-" and the "reality-principles" can be felt. But experience constantly tells us, that we are well advised to hold our breath in order to feel better in the end. Being good pays off.

Patients with frontotemporal degeneration, a type of dementia much rarer than Alzheimer's, loose interest in many areas, which demand more than just a brief effort, but they do not shy away from any faux-pas [14, 15]. This loss of interest, of respect, and a negligence regarding the effects of one's action or apathy, may occasionally lead to petty crime and rarely to more severe and fatal violations of social conventions [16]. Due to the patients' "inability to will", and their lack of insight, they cannot be held responsible for what they have done or not done. This complete loss of a future perspective and motivation is usually accompanied by a loss of interest in the individual's personal past and basically in anything else.

ELEMENTS

Neurodegenerative dementias are experiments of nature, which affect extended brain areas and their functions in a systematic manner. Such lesion models illustrate the essential importance of certain parts of the brain for certain neuropsychological functions and behaviors. It would however be a misunderstanding to assume that other parts of a healthy brain were completely uninvolved in any of these tasks.

This evidence suggests that different parts of the brain are indeed particularly relevant for certain aspects of time-related experience and behavior:

- *Past:* Areas critical for the storage of long-term memory predominantly affected in Alzheimer's dementia are the medio-temporal lobe and the temporal and parietal neocortex; further parts of the brain degenerate in more advanced stages of dementia. In the earlier stages of illness patients are not confused and they are interested in their environment and future.

- *Present:* confusional states as observed in dementia with Lewy-bodies lead to a temporary inability to store and retrieve information in an organized manner, and are usually associated with other disconcerting symptoms. Most patients are anxious and feel insecure about their surroundings and future, but they do not "get their act together". Brainstem and limbic system suffer from a double pathology, cannot provide sufficient acetyl-choline and conduct the outer layers of the central nervous system directing them towards relevant signals.

- *Future:* a degeneration of the large and evolutionarily recent fronto-subcortical loops first reduces subtle social skills and eventually leads to gross misconduct. Perception, short-term memory is intact but underused for an often times stoic consumption of television or the ingestion of junk-food. The patient shows little interest in his own past, even though long-term memory would be readily available, - but for no purpose.

As long-term and short-term memory are fundamentally different, one being the slowly but constantly changing synaptic architecture of the whole brain, the other being the fast-changing electrochemical oscillations of the complete brain in the very present, it becomes quite obvious that we will not be able to accommodate these different aspects of memory in distinct anatomical parts of the brain side by side. But this is not my immediate concern, as we have set out to collect some evidence for a critical dependence of the past, present and future on different parts of the brain (this aim is far more modest than a meticulous work-up on the cerebral underpinnings of memory). We can pen down the following neurobio-logic of time (see Table **1**).

Table 1: A neurobio-logic of the present, past and future based on the lesion models of degenerative dementias (⇓ key deficit; ↓ secondary symptom; √ intact).

	Present	Future	Past
Dementia with Lewy-bodies	⇓	↓	↓
Frontotemporal degeneration	√	⇓	↓
Alzheimer dementia	√	√	⇓

Our examples have demonstrated that confusional states - as observed in dementia with Lewy-bodies - inhibit the formation of long-term memory and preclude organized retrieval from long-term memory. In spite of a clear consciousness, patients with frontotemporal degeneration have no interest in their future and make little use of their long-term memory. Patients in the early stage of Alzheimer's dementia are primarily handicapped by their problems with declarative long-term memory, but pay attention to the present and relate to future perspectives. For a more philosophical explication of these results the keen reader is referred to Augustinus [2], who has already pronounced the importance of the present (short-term memory) and of intentions, which govern our use of other cognitive realms. Dummett ([1]; see introduction) also favored the central role of the present, with future as the runner-up, well ahead of the past.

SIDELINES

It would be rather uneconomical to assume, and uneconomical to device and run a world which constantly dissolves as soon as we look the other way. Who would have an interest in such an investment and in ourselves to test us with such elaborate means? Even if we cannot rule out that this may in principle be the case, it seems much more appropriate or reasonable not to bother too much, but to take the economical stance and accept the existence of a world which will demonstrate rather the same features and similar personnel – only slightly older – when we again look into the same direction. Then no major effort is needed need to rebuild, repaint, re-inflate and animate the whole scenery including our friends and foes. A skeptic and inductive bottom-up approach clearly bears the risk of not finishing a multiplicity of tasks on time. Deductive and reasonable top-down assumptions are economical. "God" is clearly the ultimate human top down concept, and for most believers the ultimate abstraction or rather personification of morality and rules on reward learned over their lifetimes. Humans feel better with consolation and hope, perhaps even belief. Patients with Alzheimer's dementia may forget about religious details, patients with frontotemporal degeneration cannot be bothered with ethical and theological issues, and a confused mind may experience inspiring, thrilling, sometimes mystic sensations [17]. But these attitudes and insights are always reflecting the mental functions granted and limited by human brains, whether in good health, demented or confused. One can be quite sure that man and monkey share more concepts of mind-mechanics and afterlife than both species share with nematodes, simply due to the degree of similarity and dissimilarity of their nervous systems and their comparably rich environment [18]. I may not have an appropriate taste of a worm's earthly delights, therefore its transcendental tendencies rather elude me, but its memories of blissful relaxation after strong pushes rewarded by a wholesome mouthful of mud may foreshadow bigger and better bites in its one-dimensional and shady subterranean future. Heaven would probably not be a valid concept and a nematocentric inclination would generate quite different pictures of friendly staff from outer earth.

ACKNOWLEDGEMENTS

Declared none.

CONFLICT OF INTEREST

The author(s) confirm that this chapter content has no conflict of interest.

DISCLOSURE

Part of information included in this chapter has been previously published in International Psychogeriatrics, 01 October 2008, Volume 20, Issue 05, pp 863-870.

REFERENCES

[1]   Dummett, M. (2004) Truth and the past – the metaphysics of time. Columbia University Press, New York.

[2]   Augustinus (400) Quid enim est tempus / Was ist Zeit? Confessions XI / Bekenntnisse XI. Felix Meiner, Hamburg.

[3]   Achtner, W., Kunz, S., Walter, T. (1998) Dimensions of Time – the Structures of the Time of Humans, of the World, and of God. William B. Eerdmans PC, Grand Rapids, Michigan.

[4]   World Health Organization (1991) 10th Revision of the International Classification of Diseases (ICD-10). WHO.

[5]   Tulving, E. (2002) Episodic memory: from mind to brain. Annu Rev Psychol 53: 1–25.

[6]   Levine, B. (2004) Autobiographical memory and the self in time: brain lesion effects, functional neuroanatomy, and lifespan development. Brain & Cognition 55: 54–68.

[7]   Förstl, H. (Ed.; 2006) Frontalhirn – Funktionen und Erkrankungen. 2nd Ed., Springer, Heidelberg.

[8]   James, W. (1890) The Stream of Consciousness. Psychology, Chapter XI.

[9]   Crick, F., Koch, C. (2003) A framework for consciousness. Nature Neuroscience 6: 119–126.

[10] Pöppel, E. (1997) A hierarchical model of temporal perception. Trends in Cognitive Science 1: 56–59.

[11] Aziz, V.M., Warner, N.J. (2005) Capgras syndrome of time. Psychopathology 38: 49–52.

[12] Leibniz, G.W. (1714) Monadologie.

[13] Buckner, R.L., Carroll, D.C. (2006) Self-projection and the brain. Trends in Cognitive Neuroscience 11: 49–57.

[14] Gregory, C., Lough, S., Stone, V., Erzinclioglu, S., Martin, L., Baron-Cohen, S., Hodges, J.R. (2002) Theory of mind in patients with frontal variant frontotemporal dementia and Alzheimer's disease: theoretical and practical implications. Brain 125: 752–764.

[15] Sturm, V.E., Rosen, H.J., Allison, S., Milller, B.L., Levenson, R.W. (2006) Self-conscious emotion deficits in frontotemporal lobar degeneration. Brain 129: 2508–2516.

[16] Diehl, J., Ernst, J., Krapp, S., Förstl, H., Nedopil, N., Kurz, A. (2006) Frontotemporale Demenz und delinquentes Verhalten. Fortschritte der Neurologie Psychiatrie 74: 203–210.

[17] Förstl H (Hrsg.; 2007) Theory of Mind – Neurobiologie und Psychologie sozialen Verhaltens. Springer, Heidelberg.

[18] Zentall, T.R. (2006) Mental time travel in animals – a challenging question. Behavioural Processes 72: 173–183.

Send Orders of Reprints at reprints@benthamscince.net

CHAPTER 7

Time Experience During Mystical States

Ulrich Ott[*]

Bender Institute of Neuroimaging, University of Giessen, Otto-Behaghel-Str., 10H, 35394 Giessen, Germany

Abstract: Phenomenological analyses and questionnaire studies have shown that changes in time experience are a prominent feature of mystical states of consciousness. Spiritual traditions employ a variety of methods to induce these states. Research on meditation and psychedelic drugs can help to identify involved brain mechanisms and neural correlates of mystical states. A theory is presented that explains the experience of unity and timelessness with a phase transition to extended coherent EEG gamma activity. Based on this theory, ascetic and meditative practices can be understood as rational methods to enable qualitative shifts in the large-scale organization of brain dynamics. Some supporting evidence for the theory comes from a study with Buddhist monks. Research on mystical experiences has to deal with many methodological challenges and requires a close collaboration of scientists and religious practitioners. Research of this kind can yield important insights into the relativity of reality and its relation to brain functioning.

Keywords: Time perception, mysticism, meditation, psychedelic drugs, EEG, neuroscience, brain dynamics, non-linear systems, attractor, singularity.

1. INTRODUCTION

Time plays a crucial role in our personal experience and in the way we organize, schedule our life. When we start to think about what time actually is, we quickly realize that time is a fundamental principle of how we perceive the world. It is hardly possible to imagine a "timeless" world. However, the attempt is worthwhile because it reveals the close connection between time and other basic elements of our reality, *i.e.,* space and matter. At the moment time ends, everything somehow has to freeze in our imagination. It is the movement of objects in space and the rhythmicity of processes within and around us that implies, requires, calls for the notion of time.

*Address correspondence to Ulrich Ott: Bender Institute of Neuroimaging, University of Giessen, Otto-Behaghel-Str., 10H, 35394 Giessen, Germany; Tel: +49-641-9926342; Fax: +49-641-9926309; E-mail: ulrich.ott@psychol.uni-giessen.de; ott@bion.de

Apart from thought experiments and philosophical reflections on the nature of time and its contribution to reality, the concept of time has proven useful and sometimes critical for our survival as biological organisms; the knowledge of the seasons might serve as an illustrative example. The long-term storage of information, combined with the ability to recall and to predict future events – on time-scales ranging from fractions of a second to decades –, is a remarkable feature of our brain, enabling effective motor control as well as future planning and intelligent decisions.

The disciplines of neuropsychology and cognitive neuroscience have accumulated a wealth of amazing insights regarding the implementation of different timing functions in the brain. Studies on selective impairments in patients demonstrate the distributed nature of memory functions and uncover the dependence of our personal life record on distinct brain areas, *e.g.,* the hippocampus (for details, see the contribution of H. Förstl in this volume). Time is an omnipresent factor in cognitive functions like verbal communication or causal reasoning, and in goal-directed behavior in general. Everyday life is to a large extent made up of the execution of scripts, prototypical sequences of actions succeeding one another, adapted to the actual situation with varying degrees of conscious awareness, performed in order to fulfill our needs. Successful behavior requires a sophisticated, fine-tuned synchronization of perception, cognition, and action, accompanied by corresponding activation patterns in the underlying neural circuitry.

In an ordinary awake conscious state, subjective time is usually in good agreement with physical time as measured by clocks. However, subjective time experience can be severely altered under certain conditions. Slowing and acceleration of time and the complete loss of the sense of time are common during altered states of consciousness [1, 2]. Obviously, the brain mechanisms that provide a time framework in the service of instrumental behavior can be transiently suspended. Conditions known to facilitate this process are, for instance, dream sleep, hallucinogenic drugs, and states of absorption, where all representational resources are engaged by an intensive experience [3].

In the following, it is shown that profound changes in time experience are a key feature of mystical states of consciousness. According to James [4], these states are the root and centre of personal religious experience. A theory is presented that explains features of these states and the means to induce them on the basis of the electrical brain dynamics in the gamma frequency range.

2. PHENOMENOLOGY OF MYSTICAL STATES

In psychological research, the definition of states of consciousness typically relies on the description of features. Each state can be mapped into a multi-dimensional space by rating the actual experience with reference to basic dimensions [5, 6]. Which changes in subjective experience and mental functioning are characteristic for mystical states? In his famous lecture on the varieties of religious experience, James emphasized four attributes of these states: ineffability, noetic quality[1], transiency, and passivity [4]. Similar attributes – with the exception of passivity – were also identified by Stace, who presented an elaborated list of nine features: 1. Unity, 2. Transcendence of space and time, 3. A deeply felt positive mood, 4. Sense of sacredness, 5. Objectivity and reality, 6. Paradoxicality, 7. Alleged ineffability, 8. Transiency, 9. Persisting positive changes in attitude and behavior [7].

While mystics have stressed that the nature of the mystical experience cannot be adequately conveyed by words, they have in fact provided quite detailed accounts. Therefore, Stace spoke of "alleged" ineffability. Based on the analysis of reports from different eras and cultures he came to the conclusion that all mystical experiences have certain core features in common. The importance of "unity", heading his list, is also reflected in the term unio mystica and the central insight of mystics that "all is one".

For the current discussion, the second feature is of special interest. Stace asserted that both, space and time, are profoundly altered in spontaneous or psychedelic mystical states, and he presented several quotations to support this claim. More evidence of this kind can be found in a recent book by Marshall, who compiled a comprehensive feature list including a separate entry for changes in time

---

[1] (Greek root "noos" = mind). The term was explained by James as follows: "… mystical states seem to… be also states of knowledge. They are states of insight into depths of truth unplumbed by the discursive intellect. They are illuminations, revelations, full of significance and importance, all inarticulate though they remain; and as a rule they carry with them a curious sense of authority for after-time."

experience: "Time 'stops'; past, present, future coexist; harmonious flow" [8]. Thus, reports by the mystics suggest that there is indeed a way – while living in space and time – to go beyond the confines of time and to perceive reality in a "timeless way" (see the contribution of A. Nicolaidis and the description of several mystical traditions in the contribution of W. Achtner in this volume). While the 'universal core' hypothesis proposed by Stace has been frequently criticized, his typology has proven fruitful for empirical research. Based on the analysis of Stace, Hood [9] developed the mysticism scale, a questionnaire, which has been widely used in the field of the psychology of religion. Four items of this scale address changes in the sense of time and space (example item: "I have had an experience in which I had no sense of time or space.").

International questionnaire research on the phenomenology of altered states of consciousness induced by a variety of methods [10] has also identified mystical experience as one main factor (named "oceanic boundlessness" with reference to the term "oceanic feeling" coined by Freud). Items with high loadings on this factor express feelings of eternity and unity: "I experienced past, present and future as a unity.", "It seemed to me that my environment and I were one."

In many mystical traditions, meditation is a prominent method to achieve states of mystical insight (see next section). Indeed, the phenomenological analysis of deep meditation states [11] yielded similar features to those of Stace listed above. Again, a changed sense of reality was reported by advanced meditators, characterized by timelessness and a unitary experience. Scales developed to assess meditation depth include items related to time experience, *e.g.*, "I lost all sensation of time" [12], and "The sense of time disappeared" [13]. The loss of the sense of time is frequently experienced by advanced meditators and indicative of a deep stage of meditation. However, hallmark of the deepest stage of meditation is the experience of "non-duality", where the usual subject-object dichotomy dissolves and gives way to an experience of all-encompassing unity [13].

3. TRADITIONAL METHODS OF THE MYSTICS

The doctrines of the world religions might differ considerably in many respects. Nevertheless, striking similarities exist regarding the methods employed by their mystical branches. Mysticism can be defined as "the pursuit of achieving communion or identity with, or conscious awareness of, ultimate reality, the divine, spiritual truth, or God through direct experience, intuition, or insight;... the purpose of mysticism and mystical disciplines such as meditation is to reach a state of return or re-integration to Godhead. A common theme in mysticism is that the mystic and all of reality are One" [14].

In contrast to religious belief systems, mysticism can be conceived as an empirical approach. Statements of the mystics on the nature of the human soul as being an entity beyond space and time are not based on speculation but on personal experience. The mystical seeker intends to verify the teachings by observing a conducive life-style and by practicing traditional methods of mental and spiritual training. Living in solitude as hermit or in a monastery as member of an order is believed to reduce distractions and to support an ethically sound life conduct. Ascetic withdrawal from worldly pleasures, *e.g.,* fasting and sexual abstinence, and a variety of meditation techniques have been recommended to further spiritual development.

Similarities between the methods used by different mystical traditions – cultivation of positive emotions, sitting in silence, focusing attention on heart, breath, mantras *etc.* – suggest an involvement of biological mechanisms common to all human beings. Neuroscience has just begun to investigate the neural correlates of spiritual practices and experiences with modern EEG mapping and neuroimaging techniques. Meditation stands out as a traditional method of mysticism, which can be conceptualized and studied as a form of mental training [15] with measurable effects on brain structure [16] and functioning (for review, see [17]).

As mentioned above, meditators report mystical experiences and changes in time experience. Mystics have discovered that meditation is able to induce these states. Science has to elucidate, how meditation and prayer influence brain functioning and enable such dramatic shifts in the experience of self and reality. The training of attention and the cultivation of emotional attitudes like humility, equanimity, compassion, and love (*e.g.,* with the "prayer of the heart" of orthodox Christianity) have an impact on the brain. The plasticity of the brain – the neural substrate of consciousness – opens up the possibility to experience states of mystical enlightenment, which are very different from ordinary consciousness. One should expect that the profound changes in subjective experience are mirrored at the level of brain dynamics.

Another line of evidence pointing to specific physiological mechanisms underlying mystical experiences comes from studies on psychedelic drugs. In many cultures, these drugs have been used in spiritual settings to induce altered states of consciousness. Today, we know that a receptor of the serotonergic neurotransmitter system (5-HT2a) is the common target site of all hallucinogens [18]. The similarities of natural and drug-induced mystical experiences suggest a common neural mechanism involving the serotonergic system [19, 20].

4. GLOBAL GAMMA COHERENCE AS CORRELATE OF MYSTICAL ENLIGHTENMENT

The electroencephalogram (EEG) measures the electrical activity of the brain through electrodes, which are placed at the scalp. Electrical signals recorded from a living human brain are characterized by oscillations at different frequencies. For the classification of these brain waves, the following frequency bands have been defined: delta (δ; below 4 Hz), theta (θ; 4 to 7 Hz), alpha (α; 8 to 12 Hz), beta (β; 13 to 30 Hz), and gamma (γ; above 30 Hz). In general, different levels of wakefulness are associated with certain EEG bands. During wakefulness, α and β activity dominate the EEG, whereas θ and δ activity is indicative for light and deep sleep, respectively.

The amplitude of the brain waves depends on the number of neurons firing synchronously. During relaxed wakefulness with eyes closed, strong α EEG activity (30 to 150 μV) is typically observed at occipital leads located above the visual cortex (idle rhythm, entraining many neurons). When the eyes are opened and visual input has to be processed, α activity is instantly replaced by a less synchronized pattern of β activity with much lower amplitudes (below 30 μV).

Oscillations in the γ range have been overlooked for a long time in EEG recordings due to their very low amplitudes (below 10 μV) and short duration. However, the scientific interest in γ oscillations has increased strongly within the last two decades because they appear to be closely linked to attention, perception, and cognition.

One of the first researchers ascribing a significant functional role to γ oscillations in the EEG was Sheer [21]. According to his view, γ oscillations are the neural correlates of focused arousal, which occurs in circumscribed neuron populations for short durations during perception, memory recall, and cognition. Shortly afterwards, Singer [22] proposed coherent oscillations in this frequency range as a solution for the so-called "binding problem" in perception. In the brain, a perceived object is represented by firing cell assemblies in distinct areas coding for attributes like color, movement, or shape. Representations belonging to the same object appear to fire coherently (in phase), while representations belonging to different objects do not.

The theory of perceptual binding through coherent γ oscillations was extended by Llinás *et al.* [23] to temporal binding and cognition in general. They proposed that high-frequency oscillations in two thalamo-cortical resonance loops provide content and context of cognition that arises from the temporal integration of sensory input and directed attention. Using magnetoencephalography, they observed a systematic phase shift of γ oscillations across the cortex. During wakefulness, sensory events caused a resetting of this "scan". In another experimental study [24], auditory clicks could only be differentiated if they produced a separate resetting; if one occurred shortly after the other (within less than 12 ms) they were perceived as a single event. The authors conclude that the spatially and temporally well-organized patterns of γ activity play an important role for coincidence detection, the experience of simultaneity, and the integration of multiple sensory components into unified percepts.

Assuming that phase differences between γ oscillations in the EEG play indeed a crucial role for the discrimination of objects in space and time, a transition to a state of global γ coherence could have enormous effects on conscious experience. A theory, first published in my doctoral thesis [25], predicts that a phase synchronization of γ EEG activity across the whole cortex – including the representation of the own body – should (1) result in the perception of all things as being one and (2) disrupt the time base implemented by the pace of the scan described above, leading to a loss of the ordinary time experience.

In computer simulations of the electrical brain dynamics, states of global synchrony have to be prevented actively by inhibitory processes, because they are ineligible for information processing [22]. Stimulus discrimination and the sequencing of perception, cognition, and action, require the large-scale coordination of activation patterns based on a delicate balance of neural excitation and inhibition. The subjective experience of a stream of consciousness with continuously changing contents mirrors the parallel distributed processing of information in the brain. Under normal conditions, executive functions located in the frontal cortex effectively regulate attention and behavior according to flexible scripts (see introduction). During the execution of scripts, attention is selectively focused on relevant objects only for short durations and thus γ oscillations in the EEG are maintained only for a few cycles. A transient hypofrontality could disrupt executive control and enable a variety of altered state of consciousness [26].

Up to now, only a few studies have investigated changes in γ coherence during altered states of consciousness. Stuckey *et al.* [27] found increases in global γ coherence ("widely distributed cortical hyper-coherence") during peak experiences in two subjects who took ayahuasca, a psychedelic drug. Two early studies, reporting high-amplitude γ EEG activity in meditators, were already mentioned by Sheer ([21]; for a detailed description and critique, see the unsuccessful replication by Ott [25]). Finally, Lutz *et al.* [28] published a high-quality EEG study on Tibetan monks with long-standing meditation experience. They found sustained high-amplitude γ activity and cross-hemispheric phase-synchrony during meditation (generating a state of unconditional loving-kindness and compassion). Moreover, increased levels of γ activity in the baseline EEG were correlated with the amount of meditation practice. However, γ activity during meditation showed a definite structure with centers at frontal and posterior regions, while the hypothesis of global γ synchronization would predict high levels of γ activity and coherence over the whole scalp.

5. IMPACT OF MEDITATION ON BRAIN DYNAMICS

The theory of global γ coherence can provide new insights and a rational understanding, in which way meditation techniques facilitate the occurrence of mystical states of consciousness.

First, meditation is usually practiced in a silent environment with strongly reduced external stimulation. In addition, a motionless sitting posture reduces proprioceptive input. This mild form of sensory deprivation reduces disturbances that cause perturbations of the brain dynamics, *e.g.,* startle or orienting responses. In such a quiet setting, the reduced need for inhibition can lead to an increased perceptual sensitivity and a lowered firing threshold at the neural level, which can also account for the vivid imagery reported frequently by participants of intensive meditation retreats.

In contrast to the multitasking demands of everyday life, a meditator merely has to follow the meditation instructions. However, novice practitioners soon discover that it is rather difficult to keep the attention focused on a meditation object or to maintain a state of mindfulness for prolonged periods of time. The stream of consciousness, usually governed by a great variety of needs, emotions, and task demands, includes a lot of discursive thoughts, recollections, and forward-looking imagination – symbolic operations that have to be restraint in favor of an awareness of the present sensations. At the beginning, this requires executive control functions and repeated efforts not to get lost in thoughts. At later stages, when the mind has calmed down, voluntary control and ego-centered goal-directed striving become a hindrance for deeper meditation experiences. Obviously, the transition to states of mystical union requires an attitude of letting go ("passivity") and devotion.

Buddhist meditation traditions, *e.g.,* Vipassana, focus on equanimity and transience to counteract aversions, desires, and ego-centered thinking. Conditioned emotional response patterns rooted in pleasant and unpleasant bodily sensations have to be dissolved in a process of realization and purification that leads to a state of detachment, clear perception, and mindfulness. At the neural level, inhibitory filter mechanisms could be reduced. A kind of highly excitable equilibrium state of brain dynamics could facilitate the transition to a state of global γ coherence involving endogenous resonance and standing wave phenomena. From the perspective of non-linear dynamics, this state would represent an attractor comparable to the activity during epileptic seizures or photic driving[2] (for a similar chaos-theoretical view, see the contribution of W. Achtner in this volume).

---

[2]Photic driving denotes the phenomenon that rhythmic (flickering) visual stimulation in the frequency range of the electroencephalogram can induce brain waves of the same frequency range.

Subjective experiences in such a "singularity" of consciousness, where the representations of space, time, and the individual itself collapse, would be qualitatively different from other changes in time experience during meditation caused by boredom (slowing of time) or by a lack of awareness of the passing of time due to daydreaming or mental absorption.

A detailed discussion, how other common meditation techniques, like the repetition of mantras or the cultivation of all-encompassing love and compassion, might affect brain functioning and promote γ synchronization in the EEG, goes beyond the scope of this paper. The development of a comprehensive theory of mystical states and their induction requires a lot of empirical work in the future. Some of the challenges and possible benefits of this kind of research are outlined in the concluding section.

6. FUTURE PROSPECTS

More than twenty years ago, an open-minded dialogue between contemplative scholars and scientists has been initiated by the Dalai Lama and other members of the Mind & Life Institute (http://www.mindandlife.org). The extensive scope and depth of this exchange is documented in many publications and illustrates how much both, religion and science, can profit from an exchange of ideas and insights. The study of meditation and religious experiences requires a close cooperation between researchers and practitioners based on interest and mutual appreciation. In the past, rigid doctrines and reductionist positions have hampered a fruitful interaction all too often.

Recently, nuns participated in a neuroimaging study on the neural correlates of mystical union [29]. In laboratories around the globe, effects of meditative and religious practices on body, mind, and health are examined to achieve a better understanding of the involved physiological and psychological mechanisms and to establish new clinical interventions. This kind of research has to face many methodological challenges. Relevant factors like love and devotion are difficult to measure and to control experimentally. Practitioners with long-standing experience are rare and might hesitate to participate. Moreover, bare laboratory settings as well as the measurement devices are hardly conducive for the occurrence of spiritual experiences. Finally, the intention to attain a mystical state in the course of an experiment can be detrimental because it engages executive functions that inevitably prevent this state to occur.

Insights of the great spiritual teachers gained in mystical experiences have an enormous impact on society and its ethical foundations. Contemporary mystics confirm the significance of these experiences for a deep understanding of the nature of reality. For scientists, mystical states represent an opportunity to explore the correlation between subjective reality and brain function and to realize the limitations imposed on their conception of reality by their own consciousness state. Scientists who practice the traditional methods can make personal experiences of timelessness that help to overcome these limitations.

Current models of the brain dynamics (*e.g.,* [30, 31]) explain how complex non-linear interactions lead to the transient formation of neuronal assemblies, firing synchronously at γ frequencies. Sophisticated mechanisms of large-scale integration enable the emergence of coherent cognition and behavior [32]. However, research on altered states of consciousness shows that assembly formation depends critically on intact brain structure, neurochemical balance, and average levels of arousal [2]. For this reason, natural rhythms (*e.g.,* sleep-wake-cycle), diseases (*e.g.,* schizophrenia, epilepsy), pharmacological substances (*e.g.,* anesthetics, psychedelic drugs), and psychophysiological techniques (*e.g.,* hyperventilation, meditation) that exert a strong influence on the brain dynamics can lead to the emergence of entirely different modes of cognitive functioning and time experience.

Timeless moments during mystical states remind us that indeed all subjective experiences – including such elementary features of ordinary reality like time – are ultimately based on neuronal representations generated by the brain. Future models, which take into account the many degrees of freedom of brain dynamics, should be able to describe, explain, and predict state changes induced by spiritual practices and the characteristic features of attained mystical states of consciousness.

ACKNOWLEDGEMENTS

The Bender Institute of Neuroimaging at the University of Giessen (http://www.uni-giessen.de/cms/bion) is an external research unit of the Institute for Frontier Areas of Psychology and Mental Health, Freiburg, Germany (http://www.igpp.de).

CONFLICT OF INTEREST

The author(s) confirm that this chapter content has no conflict of interest.

REFERENCES

[1]   Ludwig, A. M. (1966). Altered states of consciousness. *Archives of General Psychiatry, 15*, 225–234.

[2]   Vaitl, D., Birbaumer, N., Gruzelier, J., Jamieson, G., Kotchoubey, B., Kübler, A., Lehmann, D., Miltner, W.H.R., Ott, U., Pütz, P., Sammer, G., Strauch, I., Strehl, U., Wackermann, J. & Weiss, T. (2005). Psychobiology of altered states of consciousness. *Psychological Bulletin, 131*, 98–127.

[3]   Ott, U. (2007). States of absorption: in search of neurobiological foundations. In G. A. Jamieson (Ed.), *Hypnosis and consciousness states: the cognitive-neuroscience perspective* (pp. 257–270). New York: Oxford University Press.

[4]   James, W. (1902). Varieties of religious experience, a study in human nature. Retrieved September 20, 2010, from: http://www.gutenberg.org/etext/621.

[5]   Tart, C. T. (1980). A systems approach to altered states of consciousness. In. J. Davidson & R. Davidson (Eds.), *The psychobiology of consciousness* (pp. 243–269). New York: Plenum.

[6]   Pekala, R. J. (1991). Quantifying consciousness: an empirical approach. New York: Plenum.

[7]   Stace, W. T. (1961). *Mysticism and philosophy*. London: Macmillan.

[8]   Marshall, P. (2005). *Mystical encounters with the natural world*. Oxford: Oxford University Press.

[9]   Hood, R.W. (1975). The construction and preliminary validation of a measure of reported mystical experience. *Journal for the Scientific Study of Religion, 14*, 29–41

[10]   Dittrich A., von Arx, S. & Staub, S. (1985). International study on altered states of consciousness (ISASC). Summary of the results. *German Journal of Psychology, 9*, 319–339.

[11]   Gifford-May, D. & Thompson, N. L. (1994). 'Deep states' of meditation: phenomenological reports of experience. *Journal of Transpersonal Psychology, 26*, 117–138.

[12]   Ott, U. (2001). The EEG and the depth of meditation. *Journal for Meditation and Meditation Research, 1*, 55–68.

[13]   Piron, H. (2001). The Meditation Depth Index (MEDI) and the Meditation Depth Questionnaire (MEDEQ). *Journal for Meditation and Meditation Research, 1*, 69–92.

[14]   http://en.wikipedia.org/wiki/Mysticism

[15]   Barinaga, M. (2003). Studying the well-trained mind. *Science, 302* (October 3), 44–46.

[16]   Lazar, S. W., Kerr, C., Wasserman, R., Gray, J. R., McGarvey, M., Quinn, B. T., Dusek, J. A., Benson, H., Rauch, S. L., Moore, C. I. & Fischl, B. (2005). Meditation experience is associated with increased cortical thickness. *NeuroReport, 16*, 1893–1897.

[17]   Cahn, B. R. & Polich, J. (2006). Meditation states and traits: EEG, ERP, and neuroimaging studies. *Psychological Bulletin, 132*, 180–211.

[18]   Nichols, D. E. (2004). Hallucinogens. *Pharmacology and Therapeutics, 101*, 131–181

[19]   Goodman, N. (2002). The serotonergic system and mysticism: could LSD and the nondrug-induced mystical experience share common neural mechanisms? *Journal of Psychoactive Drugs, 34*, 263–272.

[20]   Ott, U., Reuter, M., Hennig, J. & Vaitl, D. (2005). Evidence for a common biological basis of the absorption trait, hallucinogen effects, and positive symptoms: epistasis between 5-HT2a and COMT polymorphisms. *American Journal of Medical Genetics (Neuropsychiatric Genetics), 137B*, 29–32.

[21]   Sheer, D. E. (1989). Sensory and cognitive 40-Hz event-related potentials: Behavioral correlates, brain function, and clinical application. In E. Basar, & T. H. Bullock (Eds.), *Brain dynamics: progress and perspectives* (pp. 339–374). Berlin: Springer.

[22]   Singer, W. (1993). Synchronization of cortical activity and its putative role in information processing and learning. *Annual Review of Physiology, 55*, 349–374.

[23]   Llinás, R., Ribary, U., Joliot, M., & Wang, X.-J. (1994). Content and context in temporal thalamocortical binding. In G. Buzáki, R. Llinás, W. Singer, A. Berthoz, & Y. Christen (Eds.), *Temporal coding in the brain* (pp. 251–272). Berlin: Springer.

[24]   Joliot, M., Ribary, U., & Llinás, R. (1994). Human oscillatory brain activity near 40-Hz coexists with cognitive temporal binding. *Proceedings of the National Academy of Sciences of the USA, 91*, 11748–11751.

[25]   Ott, U. (2000). *Merkmale der 40 Hz-Aktivität im EEG während Ruhe, Kopfrechnen und Meditation* [Characteristics of 40 Hz-EEG-activity during rest, mental arithmetic, and meditation] (Schriften zur Meditation und Meditationsforschung, Band 3 [Writings on meditation and meditation research, vol. 3]). Frankfurt: Peter Lang.

[26]   Dietrich, A. (2003). Functional neuroanatomy of altered states of consciousness: the transient hypofrontality hypothesis. *Consciousness and Cognition, 12*, 231–256.

[27]   Stuckey, D. E., Lawson, R., & Luna, L. E. (2005). EEG gamma coherence and other correlates of subjective reports during ayahuasca experiences. *Journal of Psychoactive Drugs, 37*(2), 163–178.

[28]   Lutz, A., Greischar, L.L., Rawlings, N.B., Ricard, M. & Davidson, R.J. (2004). Long-term meditators self-induce high-amplitude gamma synchrony during mental practice. *Proceedings of the National Academy of Sciences, 101*(46), 16369–16373.

[29]   Beauregard, M. & Paquette, V. (2006). Neural correlates of a mystical experience in Carmelite nuns. *Neuroscience Letters, 405*, 186–190.

[30]   Bojak, I. & Liley, D. T. J. (2007). Self-organized 40 Hz synchronization in a physiological theory of EEG. *Neurocomputing, 70*, 2085–2090.

[31]   Freeman, W.J., & Vitiello, G. (2006). Nonlinear brain dynamics as macroscopic manifestation of underlying many-body field dynamics. *Physics of Life Reviews, 3*, 93–118.

[32]   Varela, F. J., Lachaux, J.-P., Rodriguez, E., & Martinerie, J. (2001). The brainweb: phase synchronization and large-scale integration. *Nature Reviews Neuroscience, 2*, 229–239.

CHAPTER 8

Forms of Time: Unity in Plurality

Jiří Wackermann[*]

Department of Empirical and Analytical Psychophysics, Institute for Frontier Areas of Psychology and Mental Health, Freiburg i. Br., Germany

Abstract: In the present paper we examine different forms of time, focusing especially on the relation between the objective time and time of subjective experience. Experimental data from psychophysical studies on duration perception are presented, and the 'dual klepsydra model' of internal time representation is introduced. Relevant notions of abstract chronometry are briefly reviewed, and properties of time-scales generated by 'klepsydraic clocks' are studied. It is shown that these time-scales allow for a consistent time-keeping, although they are non-uniform with respect to the objective time. Parallels between our findings and E.A. Milne's theory, revealing plurality of cosmological time-scales, are pointed out. Unitary and uniform time-scale then appears as an intersubjective construct, arising from communication of shared phenomenal fields by an ensemble of observers. Limits of the presented approach, encountered at the asymptote of cosmological and theological thought, are drawn.

Keywords: Chronometry, clock, cosmic time, dual klepsydra model, duration discrimination, duration reproduction, horological function, klepsydraic clock, klepsydraic reproduction function, Milne's cosmology, perception of time, psychophysics, subjective time, time scale.

1. TIME, THE MEASURE AND THE MEASURABLE

Time is not a thing among other things; time is rather a form of the process of universal change, given in a succession of distinct world's states. The world is a theatre of appearances arising from and receding to an undifferentiated background. Things of the world are in a continuous flux, in permanent alternation and succession, according to the order of time [1]. In this archaic view, time was a synonym for the ordering principle ruling the Universe, κόσμος. Time

[*]**Address correspondence to Jiří Wackermann:** Department of Empirical and Analytical Psychophysics, Institute for Frontier Areas of Psychology and Mental Health, Wilhelmstr. 3a, D-79098 Freiburg i. Br., Germany; Tel: +49 761 2072171; Fax: +49 761 2072199; E-mail: jw@igpp.de

was primarily cosmic time, manifesting itself in the regularity of the world's process. In Plato's philosophical myth, time was a "moving image of eternity" [2], constructed on mathematical principles (fixed arithmetic ratios). Therefore, time can be conceived as a quantitative measure of change, or, in Aristotle's classical definition, the "number of motion in respect of 'before and after' " [3].

This notion of arithmeticized time was a precursor of the concept of 'mathematical time', which was enforced by the rise of the new, dynamical physics (Galilei, Huygens, Newton *et al.*). Conceptually, time in physics is a parameter of dynamic equations, increasing monotonically with the succession of world states. Operationally, time is measured by a clock, that is, a physical system implementing a locally bound inertial motion and thus believed to provide a uniform measure of time (Huygens). Newton postulated "absolute, true, mathematical time (which) flows equably without relation to anything external" [4]: an objective reality independent from clocks and observers. Later criticism of Newton's concepts (Mach, Poincaré; see Section 2) emphasized the relational notion of time, and motivated operational analysis of two key notions, namely, the simultaneity of events, and the uniformity of time-scales: a 'prelude' to Einstein's special relativity and subsequent theoretical developments.

There is a peculiar dialectic in the historical movement of the concept of time: first, the cosmic ideal of uniform time was projected upon local clocks, to be later re-constructed out of scattered clocks' readings [5]. And more, a residue of this development is the ambiguity permeating our language and thought: on the one hand, time is conceived of as the measure of all things but, on the other hand, time itself is thought of as some-thing measurable.

2. REALITY OF TIME: OBSERVATIONAL APPROACH

It is common to speak of 'flow of time'; our language thus conserves a highly pictorial metaphor of time as something moving in itself, something stuff-like. Clocks, then, can be seen as instruments capable to detect, measure, and indicate the amount of 'the flowed', that is, elapsed duration. In this naïve view, clocks are like spinning windmills driven by the 'wind of time': maybe a nice poetical image, but certainly a wrong concept [6]. The misleading metaphor is further supported by some authors' claims that 'passage of time' is an immediate, introspectively accessible datum of consciousness [7].

While this popular, intuitive view fits well with the image suggested by Newton's concept of "absolute time" unrelated to any empirical reality or observable processes, it was rejected by Ernst Mach "as an idle metaphysical conception". Mach pointed out that "time independent of change [...] absolute time can be measured by comparison with no motion [...] has therefore neither a practical nor a scientific value" [8]. In his view, it is incorrect to say that we "measure the changes of things by time. Quite the contrary, time is an abstraction, at which we arrive by means of the changes of things; made because we are not restricted to any one definite measure, all being interconnected" [9].

In a similar vein, Henri Poincaré emphasized the multiplicity of possible operational definitions of a measure of time, and concluded: "Time should be so defined that the equations of mechanics may be as simple as possible [...]. [T]here is not one way of measuring time more true than another; that which is generally adopted is only more convenient" [10]. Poincaré thought primarily of simple expression of laws of mechanics (in Newtonian tradition) [11], but Mach properly observed that any physical process may serve as the basis of time measure—for example, "the excess of the temperature of a cooling body over that of its surroundings" [12]. In fact, there are no grounds to prefer a mechanical motion as a time-keeping standard. Moreover, there is no a priori necessity that time-scales conveniently applicable in different phenomenal domains be identical [13].

In a brief summary: Mach's and Poincaré's critiques amounted to a complete revision of the concept of time; from the instrumentalist point of view, 'time is nothing else than what the clock indicates'. In this view, clocks are no more passive devices driven by time, but rather human artifacts superimposed on the phenomenal field, and actively generating a measure of time: that is, providing indications relatively to which any particular process can be compared. In observational terms, there is no 'true time', approximated by clocks, but only classes of clocks defining (in a sense to be specified later) the same time-scale. Physical theories, or individual chapters of physics (mechanics, thermodynamics, *etc.*) are self-consistent (as far as possible) descriptions of observational relations between phenomena under study; out of this relational nexus, the time dimension is, speaking with Mach, "obtained by abstraction".

This radically observational stance may be confusing for adherents of naïve realism, but also frustrating for philosophers who (mis)interpret this view as a philosophical doctrine. Indeed, the operationalist (instrumentalist, conventionalist) position is often presented (and criticized or ridiculed) as an opposite of the 'scientifically informed realism'. We should emphasize that such questions as 'does time really exist?', and arguments in favor of a positive or negative answer, are perfectly irrelevant for our investigations. As has been said in the introduction: "time is not a thing among other things". In other words, time is 'no-thing'—which of course is not the same as saying 'time is nothing,' that is, not real, inexistent. Time is as real as other abstractions are real: they are no things, no substances, but rather names of fill-in entities in a theory, co-ordinating terms in a given descriptive system.

3. PLURALITY OF TIMES?

The periodicity of the Earth's rotation served for centuries as the standard for synchronization and correction of terrestrial, mechanical clocks. At present, the measure of time is defined *via* a frequency standard based on atomic spectrometry, which is in good empirical agreement with the earlier definition [14], and serves as the basis of the 'universal time' (UT) implemented in a planet-wide system of synchronized clocks [15]. The notion of time derived from fundamental laws of physics permeates all domains of empirical reality, in sciences and technology as well as in practical, social and economical life. Due to this unitary concept, all temporal data—ranging from life-times of elementary particles or firing rates of neural cells up to the dating of historical events and geological epochs—are projected onto one unique axis of numeric coordinates.

It is often felt that this notion of uniform time is alien to the proper dynamics of biological, psychological, or social processes. It is claimed that time of physics is not intrinsically or naturally related to the 'proper time' (eigentime) of biological or social systems. So, for example, the notion of 'physiological' time has been coined in biological sciences: the proper time-scale of an individual organism's life-span, non-linearly related to physical time, and measured by the growth rate of cellular tissues [16], or *via* dynamic parameters of the organismic metabolism [17]. Moving from individual organisms to species, the eigentime of biological evolution could be defined by the evolutionary process itself, for example, by the production rate of genomic combinations. And, still in the same vein, reality of 'psychological' time, which is relatively independent from the objective 'clock time,' has been also claimed (*cf.* Section 4).

The alleged multitude of 'times' arising in diverse realms of reality evidently presents a challenge for the theoretical thought, which is directed to unitary concepts behind the diversity of phenomena. In fact, there is only one universe, one universal nexus connecting phenomena, to be understood in terms of a small number of fundamental principles and 'laws'. This ideal has been achieved, in a remarkable extent, in physics; but the 'physical reality' is merely one particular aspect of the totality of being [18]. How can the notion of different 'times', specific for other domains of experienced reality (biological, social, *etc.*), be reconciled with the idea of a unitary temporal order? Here the concept of emergence may be considered.

'Emergence' means that at certain levels of complexity of a system, apparently 'new' properties occur which do not exist in its elementary constituents. The classical example: a drop of water can be described in terms of 'macroscopic' properties—viscosity, light refraction index, *etc.*—that are undefined for single molecules of water. The original motivation for the concept of 'emergence' was to account for occurrence of life and mind, phenomena presumably irreducible to physical and chemical mechanisms. In the contemporary philosophy, 'emergence' is understood as an "epistemological, not metaphysical, category" [19]. Nonetheless, emergence is clearly an attempt to 'save appearances' (σώζειν τα φαινόμενα) from the 'nothing but'-reduction to a lower level of description— without questioning the fundamental rôle of the lower-level! This is why emergentist strategies are presently embraced by theorists searching for a non-reductionist, yet 'scientifically supported' world-view. Then, if we take the claims of 'regional times' seriously, the concept of emergence seems attractive: the low-level (molecular, cellular) processes evolve in physical time, while the higher-level (organismic) processes take place in 'biological' time, *etc.*

Yet, there are doubts, a great deal of. We have said above: time is not a thing among other things. Time is not a 'substance' bearing some 'accidents' or 'attributes'; all talk of 'properties' of time(s) is suspicious. By the same token, time is not an instance enforcing or causing something to happen. Consider, for example, the popular argument for the reality of 'biological time': an individual organism develops at a higher 'tempo' in the early than in later stages of its life-time. Indeed, a graph of average frequency of cell divisions as a function of physical time, t, reveals a non-linear relation. But this is how the development takes place, not why; it is not that there are 'two times' flowing in parallel, one of them being external and alien to the observed phenomenon, the other one causally involved in the phenomenon. The same can be said, *mutatis mutandis*, about claims of reality of 'psychological time', 'social time', *etc.* In our view, the unity of the world implies only one world process, involving all observable phenomena, in whatever domain they occur [20].

Briefly: *time is not a cause, but the form of the world process.* We do not believe in different 'kinds' of time emerging in different ontic domains. Instead, we will admit that different 'forms' of time may be needed to coordinate phenomena occurring in these domains; that is, different equivalence classes of 'clocks' based on the laws ruling the respective regions, defining different time-scales. This programmatic thesis will be exemplified in Sections 5 and 6 by our studies of the metric of 'subjective' time.

4. OBJECTIVE *VS.* SUBJECTIVE TIME

In the beginning of our investigations, we have characterized time as the aspect of the universe, which is recognized in the succession of world states. The observer does not play any special rôle in this account. The temporal order is objective: any two observers sharing a given phenomenal field [21] necessarily agree about the order of any two particular events, *e.g., 'A* precedes *B'.* Judgments about order of events are usually unambiguous [22], while our judgments about time elapsed (duration) between events are often flawed.

Our everyday lives provide rich evidence that our subjective perception of temporal durations are notoriously unreliable [23]. In a competition with a clock, we are always the losers. Our judgments of objectively equal durations may greatly vary, depending on our attentional (emotional, *etc.*) states. These variations are popularly expressed as "time runs fast," or "time goes slowly with me". It is generally agreed that "the time of the physicist does not coïncide with the system of time-sensations (*Zeitempfindungen*) [24]. As a consequence, some authors proposed that 'subjective', 'mental' or 'psychological' time is (relatively) independent from the 'objective,' 'physical' time [25]. Schematically, mental time is the time of the inner life, in which our perceptions, thoughts, *etc.* take place, just as the external world's events take place in physical time. An illustrative example from a recent study on the 'morphology of time':

> "Turning our attention away from the external clock and observing us ourselves, we discover in us quite different temporal orders than the uniform, monotonic periodicity of a pendulum or of caesium atom vibrations. We discover in us experiences of time dilatation and time contraction, which may even reach up to a cessation of time, annihilation of time" [26].

This excerpt aptly illustrates a problem with the 'subjective evidence' for the alleged reality of mental time: what does it really mean, 'in us'? On a second thought: "turning away from the clock," we do not find any evidence for time's 'speeding up' or 'slowing down'; these judgments come only *post hoc*, by comparison between the expected and actual readings of an external clock [27]. Reports from so-called altered states of consciousness, referring to time 'standing still,' 'eternal now,' *nunc stans*, are even more spectacular [28]. But these reports may serve as well as arguments against the claimed reality of mental time: they simply show that, due to the lack of an intrinsic 'sense of time' and in absence of a regular update from an external time reference, the knowledge of time data, and thus the impression of 'time passage', are lost.

However, we may dismiss the speculative concept of 'psychological time' as based on dualist presumptions, while preserving the distinction between objective and subjective time as meaningful and useful. We may, for example, ask whether we have a sort of 'sense for temporal duration'—even if a very primitive and unreliable one—underlying our subjective judgments of duration of processes in the objective world. This question is not only of interest for empirical psychology;

it also has bearing on the problem of arbitrariness of time-scales, touched in Section 2. Mach maintained that a physical process serving as a measure of time should "proceed in almost parallel correspondence with our sensation of time" [29]. Poincaré held an opposite view: "We have not a direct intuition of the equality of two intervals of time. The persons who believe they possess this intuition are dupes of an illusion" [30]. This is no more a metaphysical question; it is an empirical question to be adressed by psychophysical experiments.

5. PSYCHOPHYSICS OF TIME PERCEPTION

5.1. Experimental Data

Of interest are here psychophysical [31] data on discrimination and/or reproduction of temporal durations. In sensory psychophysics, discrimination is understood as the subject's ability to distinguish stimuli of different physical magnitudes, *e.g.*, two tones of different frequency, or two light flashes of different radiant energy. By analogy, in our experiments pairs of subsequent stimuli (*e.g.*, visual signals) of varied durations s_1 and s_2 are presented to the observers, whose task is to indicate which of the two stimuli was perceived as shorter or longer.

These experiments reveal a remarkable asymmetry in time perception. In a duration discrimination study [32] using time intervals of average duration ≈ 6 s, the ratio of subjectively indifferent durations (PSE) [33] was $s_2/s_1 \approx 0.76$; this means that the two intervals were subjectively perceived as equal if the second duration was about 3/4 of the first. In another study, using intervals of average duration of ≈ 3 s, the ratio s_2/s_1 at the PSE was ≈ 0.9: a smaller but still significant asymmetry [34]. Past durations are underestimated relatively to recently perceived ones; this indicates that the subjects' internal representation of duration changes as a function of elapsed time [35].

This effect can be observed also in experiments on duration reproduction (Fig. **1a**). A time interval of duration s is presented to the observer *via* a sensory stimulus; after a short pause of duration w, the stimulus is re-displayed; the observer's task is to indicate, *e.g.*, by pressing a response key, the moment at which the second appearance equals the first. The mean response time $\bar{r}$ (averaged over repeated trials with the same duration s) gives an estimate of the

PSE, $(s, \bar{r})$. A 'reproduction curve' is obtained by varying durations s in a certain range and plotting $\bar{r}$ as a function of s. The empirical reproduction curves typically show a progressive shortening of the response time r with increasing stimulus duration s, as seen in Fig. **1b** [36]. This negative curvature is a robust, reproducible phenomenon, found also in data reported by other researchers, but surprisingly little attention has been paid to this effect. Data from duration discrimination and reproduction experiments thus provide converging evidence for a non-conservative, 'lossy' internal representation of past time intervals—a fact to be accounted for by models of time perception.

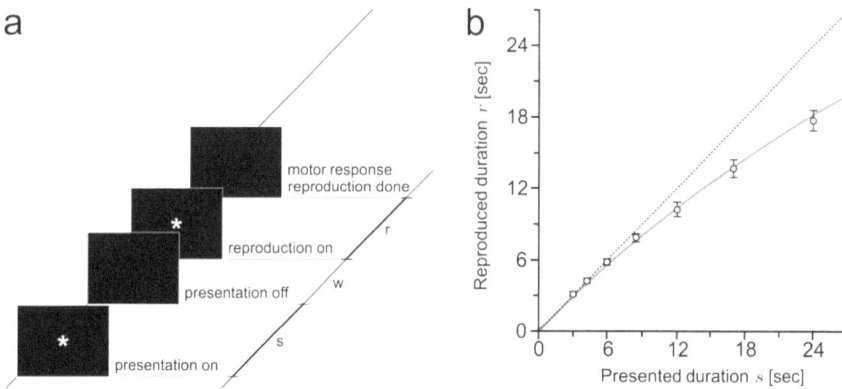

Figure 1: Time chart of the duration reproduction task (a) and a typical reproduction curve (b). Data shown in section (b) are grand means (±1 st. dev. of the mean) for a group of 12 subjects; individual means were obtained by averaging responses over 20–30 trials for each subject and duration s. The grey curve is a model-based reproduction function matched to the data (Section 5, eq. (2)).

5.2. Dual Klepsydra Model

The 'dual klepsydra model' (DKM) [37] of temporal reproduction consists of two inflow–outflow units (IOU), also called 'leaky klepsydrae' [38]. A linear IOU is described by an ordinary differential equation:

$$\frac{dy}{dt} = i - \kappa y \tag{1}$$

where $\kappa \geq 0$ is a given constant, and i is the influx rate. The influx to an IOU is a piecewise constant function, 'on' $(i > 0)$ during perception of a time interval, and 'off' $(i = 0)$ otherwise; therefore, the state y is a piecewise exponential function of time t. However, the accumulated states are never directly observed; only results of comparisons between two different IOUs enter the subject's consciousness [39].

The two IOUs are associated with perception of the two intervals (Fig. **1a**): klepsydra 1 is filled with influx i_1 during the interval $0 \leq t \leq s$, klepsydra 2 is filled with influx i_2 from time $s + w$ on, and the two klepsydrae are continuously compared. Equality of states $y_1 = y_2$ results in subjective experience of equality between respective durations, and elicits the subject's response at time $t = s + w + r$. Then the reproduced duration r depends on the stimulus duration s and the delay time w; the dependence is described by so-called 'klepsydraic reproduction function' (KRF),

$$r = \mathrm{krf}(s,w) := \kappa^{-1} \ln(1 + \eta \, (1 - e^{-\kappa s}) \, e^{-\kappa w}), \tag{2}$$

where κ is the loss rate factor in eq. (1) and $\eta = i_1/i_2$ is the influx ratio. In experiments with physically homogeneous stimuli we postulate $i_1 = i_2$, and thus $\eta = 1$. Given a collection of stimulus–response data (s_n, r_n) from an experiment with a constant delay w, we can fit the partial function $r_s(s) := \mathrm{krf}(s,w)$ to the data, adjusting the parameter κ. The partial KRF matches experimental data with very good accuracy, as shown in Fig. **1b:** the grey curve is the matched model-based reproduction curve for $w = 1.5$ s and $\kappa = 0.013$ s^{-1}. The estimates κ from individual or group-averaged reproduction data are usually in the range from 1 to 3×10^{-2} s^{-1}.

Originally, the model has been defined in abstract, physically unspecific terms. In a neurobiological interpretation, the 'influx' can be a flow of excitatory activity directed to a local neuronal assembly, and the state variable y would correspond to the mean excitation level of the assembly [40]. Relaxation times of the neural accumulators, defined by the inverse values κ^{-1}, are in the range $\approx$30 to 100 s. This value sets a natural limit to temporal discrimination of the DKM, as can be seen from the asymptotic behavior of the KRF: for $s \rightarrow \infty$, the response time r approaches the upper-bound $r_\infty := \kappa^{-1} \ln 2$. This is in agreement with empirical estimates of the region of subjectively experienced durations, which extends from a few seconds up to approx. one minute [41]; hence we call the upper-bound of the KRF the 'reproducibility horizon' of subjective time [42].

6. KLEPSYDRAIC TIME SCALES AND THE METRIC OF SUBJECTIVE TIME

The 'klepsydraic reproduction function' (KRF) introduced in the preceding section is not merely a convenient data fitting formula; it shows some outstanding analytical properties [43], strongly favoring the KRF as a model of subjective time measure. In this section we discuss time-scales generated by clocks based on the DKM principle.

First, we will briefly review some elementary concepts of abstract chronometry. — We call clock any device generating series of observably delimited periods according to some internal rule; in a terse phrase, a clock is 'a memory for a duration unit'. The number of clock periods needed to cover a given time interval is the measure of its duration with regard to the given clock. — The same principle is known from the space domain: to measure the distance AB, we translate a unit or 'standard' length u (*e.g.,* a yard-stick) by its own length along the line AB. If the distance is covered by n units, the length is determined by $(n-1)$ $u < AB \leq nu$ [44]. — In the Huygensian clock, the clock periods are marked by observable events ('ticks'), generated by a conveniently chosen physical process (*e.g.,* a swinging pendulum, a vibrating quartz crystal) and counted by an auxiliary mechanism.

Further, we need some (minimal) formalism. Let in the following t denote a chosen time reference, *e.g.,* Universal Time (UT) [45]. A clock begins its operation at an instant t_0, called the clock's 'epoch', and generates a series of 'ticks', occurring at times $t_1 < t_2 < \ldots < t_n < \ldots$. The clock's operational rule is formalized by the concept of *horological function*; this is a continuous and monotonically increasing (hence invertible) function ϑ such that $\vartheta(t_n - t_0) = n$. The clock's ticks are thus determined by the inverse function, $t_n = t_0 + \vartheta^{-1}(n)$. Two clocks, 1 and 2, are said *similar* if their horological functions, ϑ_1 and ϑ_2, are 'of the same form', that is, each one can be transformed into the other by linear scaling. 'Time' is nothing but a name for an equivalence class of similar clocks (instrumental definition). To allow for transformations between clocks' readings not only by a change of the unit, but also by a change of the epoch, we define a stronger (*i.e.,* more restrictive) relation, *linear congruence*. Clocks 1 and 2 are said to be congruent if their readings $T_j := \vartheta_j(t)$ ($j = 1,2$) are at any t related by the equation:

$$T_1 = \lambda\, T_2 + \mu \tag{3}$$

(with λ, μ being constant). A time-scale defined by an equivalence class of congruent clocks permits all elementary arithmetical operations on clock readings.

In accord with the observational principle (Section 2), a clock's readings can be compared only with another clock's readings—there is no question of a 'true time'. In operational terms, there are no 'right' or 'wrong' time-scales; trivially, any clock indicates 'its own time' correctly. There may be, however, useful, less useful, or useless time-scales. An important criterion of metric usefulness is that measurements of the same object yield identical values independently from the actual origin of the scale: in other words, the scale should be *uniform* [46]. This criterion precedes Poincaré's requirement of external simplicity: before we are concerned with the expression of the 'laws of mechanics' (or of any other phenomenal domain of interest), the clock must first allow for a simple expression of its own internal rule. This is where the notion of 'reproduction function' comes handy, as seen in the following.

In our operational definition, a clock is nothing else but a unit-reproduction device. Of interest is now the time-scale generated by a 'klepsydraic clock' (KC), operating on the DKM principle with parameters $\kappa > 0$, $\eta = 1$, and described by the reproduction function (2). Starting with a 'unit' u, the clock produces a series of periods $r_n = \mathrm{krf}(r_{n-1}, 0) = r_s^n(u)$ for $n = 1,2,\dots$. Setting for convenience $t_0 = 0$, and writing for brevity $\varepsilon := 1 - e^{-\kappa u}$, the total duration covered by n periods is:

$$t_n := \kappa^{-1} \ln(1 + n\varepsilon)$$

Therefore, the horological function of the KC is:

$$\mathscr{H}(t) = \varepsilon^{-1}(e^{\kappa t} - 1).$$

Importantly, the KC is 'reproductively consistent' [47]: this is to say that reproduction of the original unit u after a time delay t_n gives $r_{n+1} = t_{n+1} - t_n$, and so, for any $n > 0$,

$$\text{krf}(u, t_n) = \text{krf}(r_n, 0). \tag{4}$$

Or, more generally, for any $k \leq m \leq n$,

$$\text{krf}(t_m - t_k, t_n - t_m) = t_{n+m-k} - t_n, \tag{5}$$

of which eq. (4) is a special case for $k = -1$, $m = 0$. This means that the KC operates 'consistently with its own past', although its subsequent periods are not identical [48].

Further we define a 'chronometric function',

$$\chi(s, t, v) := \vartheta(t + v - s) - \vartheta(t - s),$$

where s is the clock's epoch, t is a time coordinate, and v a duration with respect to t. Consider an interval AB measured by two KCs, started at epochs $s_1 \neq s_2$ and running in parallel. The clocks will give different measures of AB, but the 'chronometric ratio'

$$\lambda := \frac{\chi(s_1, t_A, t_B - t_A)}{\chi(s_2, t_A, t_B - t_A)} \tag{6}$$

is invariant under a shift $AB \rightarrow A'B'$ [49]. Therefore, the clocks' readings $T_j := \vartheta(t - s_j)$ $(j = 1,2)$ at any instant $t \geq \max(s_1, s_2)$ are related by the equation $T_1 = \lambda\, T_2 + \mu$, where λ is given by eq. (6) and μ is another constant, as shown in Fig. **2** [50]. This is easy to generalize to an ensemble of KCs with unequal units, started at different epochs. Simultaneous KC's readings are related by equations

$$T_i = \mathsf{A}_{ij}(T_j) := \lambda_{ij}\, T_j + \mu_{ij} \tag{7}$$

and thus congruent in the sense of the above-given definition (3). Technically speaking, transforms A_{ij} form a group. In a non-technical summary: An ensemble of 'klepsydraic clocks' is capable of consistent time-keeping, and defines in this sense a 'quasi-uniform' time-scale: the same transformation rules apply [51] as for the clocks defining the uniform and universal time-scale t.

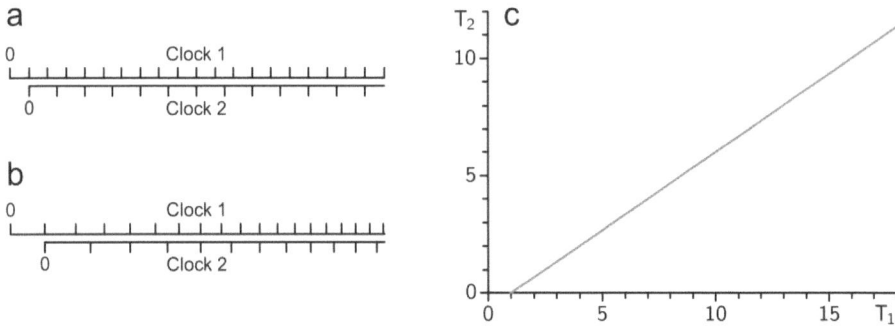

Figure 2: Tick series produced by two clocks running in parallel. (a) Uniform time-scale. (b) Klepsydraic time-scale ($\kappa = 0.1$). The clock readings are in both cases (a) and (b) transformed by eq. 7 with $\lambda_{12} = 3/2$ (constant chronometric ratio) and $\mu_{12} = 1$.

As shown in Section 5, the 'klepsydraic reproduction function' provides a good model for temporal reproduction and discrimination in human subjects. If human time perception operates on klepsydraic time-scales then an ensemble of individuals, each one equipped with a private klepsydraic clock, will share a 'quasi-uniform' time [52], which is a necessary condition for inter-subjective synchronization in concerted activities, such as cooperative work, sports, games, *etc.* A limit of this inter-subjective agreement is set by the 'reproducibility horizon' (see Section 5), implied by the klepsydraic mechanism. To maintain an agreement on temporal extensions beyond this limit, an external clock is needed. The so-called 'objective' time then appears as a convenient and physically realizable construct, extending the domain of inter-subjective agreement beyond its 'natural' horizon [53].

7. TOWARDS A UNITY OF TIME

We find unexpectedly an interesting parallel to our investigations in E.A. Milne's studies of the world's structure. Milne proposed a theory of idealized universe, building up on his 'cosmological principle': The universe should appear to its observers in such a way that the observers arrive at the same description of its contents; in other words, they find identical 'laws of Nature'. Of special interest is that part of his theory, which deals with the co-ordination of private times between individual observers, and with the construction of a commonly shared time-scale.

Milne and Whitrow's analysis [54] begins with a minimal set of assumptions: individual observers are equipped with 'arbitrarily graduated clocks'—*i.e.*, a private method of assigning temporal indices to events—and communicate pair-wise their clock readings by means of light signals. On the basis of exchanged data, observers A, B construct—each for himself—a transformation function to apply to his own and the other observer's clock readings. Due to this 'clock re-graduation', the observers A and B possess the same time-scale; symbolically, $A \equiv B$. Conditions of transitivity for the congruence between observers are found, so that observers in possession of re-graduated clocks constitute an *equivalence*, $A \equiv B \equiv C$.... Importantly, the entire theory is built upon time data given by the re-graduated clocks; spatial distances and, eventually, relative velocities between observers are obtained from light-signal transfer delays.

On this theoretical ground, the distribution of matter in the universe is studied. For this purpose, the observers are identified with material particles, and the *substratum* of the universe is defined as "an equivalence in which the density-distribution in space and time has the same description as viewed by every member" [55]. From further analysis two outstanding time-scales arise, corresponding to two types of motions of the substratum: the t-scale, on which the particles of the substratum are in uniform relative motions ('expanding universe'), and the τ-scale in which the substratum appears stationary. The two time-scales are related by a logarithmic transformation, $\tau = \log(t/t_0) + t_0$, where t_0 is the observer's present epoch.

In this way, a unitary, cosmic time is obtained: "the universe is its own clock". Yet, this universal time is by far different from the absolute time of Newtonian physics: while Newton's 'true time' was unrelated to anything external, here the universal time is intimately linked to the "actual distribution of matter-in-motion existing in the universe" [56]. Answers to fundamental cosmological questions—is the universe expanding or stationary? finite or infinite? *etc.*—then depend on which time-scale is adopted: "The spectator of the universe can take his choice. It is the interplay of these two time scales which gives the universe its interest" [57].

In Milne's interpretation, "the existence of the two scales is closely connected with the disjunction between matter and light" in the universe, so that "the τ-scale is appropriate to material particles and their dynamics and electrodynamics; the t-scale, to radiation". Even if the difference between the two time-scales is practically unimportant, still "contemporary physics has an ambiguity running through it, inasmuch it confuses the time-variables used in two distinct domains of investigations" [58]. A possible plurality of 'regional' time-scales thus remains an open, but perfectly legitimate question. As G.J. Whitrow pointed out, the implicit assumption made in physics that "all natural phenomena suitable for use as clocks must keep the same time," may not hold: "there might be natural phenomena defining physically significant time-scales which are non-uniform when referred to the rotating Earth" [59].

This brief *résumé* of Milne's theory should be sufficient to demonstrate the parallelism with our psychophysical studies of subjective time (Section 6). There we found that the time-scales generated by the observers' private 'klepsydraic clocks' allowed for a subjectively or intersubjectively consistent time-keeping. However, the limited horizon of this primitive time-keeping mechanism requires that the observers agree upon another, physically realized, unitary and uniform time, allowing synchronization of their activities across larger temporal distances. Paraphrasing Milne's words, we could say: it is the interplay of the objective, physical t-scale and the klepsydraic scales what gives the problem of 'subjective time' its interest.

If we hesitate to say that this objective, public, and unitary time 'emerges' from the subjective, private, and particular 'times', it is because we do not wish to overstretch the concept of emergence. Here we have not two realms of reality to be connected; there is only one phenomenal field shared by the observers; and the form of time the observers agree upon is just the simplest (the most economical, speaking with Mach) among others. We tend to conceive this epistemic situation from a culturalist or constructivist perspective, rather than from a naturalist and emergentist one.

8. OUTLOOK: UNIVERSALITY AND HISTORICITY

Natural science searches for regularity and invariance in the diversity of phenomena. Ideally, science makes the variety of actual or possible experience understandable by a limited number of principles and laws. Since the beginnings of modern physics (Galileo, Newton), its method consists in separation of circumstances of an observed phenomenon into potentially variable initial conditions and invariant, lawful functional dependencies; whereas really unique circumstances are completely eliminated [60]. Physics is about the repeatable and reproducible core of the phenomenal variety; all the rest—which, in fact, makes a greater deal of human participation in the world—is history.

However, the world as a primary, brute fact *is unique*: the universe is given only once [61]. The uniqueness of the world sets an asymptotic boundary where the naturalist and historical accounts of phenomena approach each other. Things happen as they happen; the history of the world is *not explained* by the universal lawful nexus but rather *identical* with it. In cosmology, physics and history of the universe are united. If we extrapolate this thought to the unity of the ideal and of the material content of the world, we are approaching the idea of creation [62].

The totality of the world implies its total unity, and thus also unity of its 'proper time'. What does it mean? We have seen in the introductory sections how the archaic view of cosmic time has been overruled by the modern, operationalist concept: 'time is what the clock indicates'. This concept was favourable for the development of positive science, from which all mythological or metaphysical residues are eliminated—but is not much useful if we think of the universe *in toto*. To have an operational clock, we have to isolate a part of the universe, and make its observations the measure of the rest of the universe; this principle plainly contradicts the idea of a total unity.

This difficulty is overcome by taking the observational principle seriously, in its strongest form: the world is a system of partial, perspectival, but *communicable* phenomenal fields, allowing for a consistent description by its scattered observers. As seen in the preceding section, this latter postulate of *equivalence of observers* was of crucial importance for the Milne–Whitrow's theory. And so, to give the phrase "the universe is its own clock" full meaning, we have to say, "the observed and communicated universe" [63]. We have also seen that this unique 'world-clock' can be still re-graduated to different time-scales: the 'ephemeral' τ-time, which is essentially a-historical, and the 'kinematic' t-time, with an origin at $t = 0$. We could say that the Milne–Whitrow's theory begins at the initial *arbitrariness* of graduation of private clocks, which is then reduced to a unitary scheme by the necessity of consistent description; but in the outcome the theory still permits some degrees of *bounded freedom*. Ultimately, the plurality of possible descriptions is an essential feature of the unitary world-time.

With these concluding reflections we are approaching—yet not transgressing—the limits of positive science. E.A. Milne's once wrote, "whilst the physicist must make clear to himself, *qua* physicist, just what his limits are, so in a larger synthesis one must tread outside these limits" [64]. If we stop at the limits, it is not with a bold claim that 'there's nothing out there, beyond the frontier'. It is rather for the respect for what is there, beyond the limit, in the realm where our 'positive' knowledge stops and has to be superseded by an understanding of a different order.

ACKNOWLEDGEMENTS

The author wishes to thank Argyris Nicolaidis and Wolfgang Achtner for the invitation to participate in the SR21 conferences and to contribute to the present volume.

CONFLICT OF INTEREST

The author confirms that this chapter content has no conflict of interest.

REFERENCES

[1]   Anaximander: "κατά την του χρόνου τάξιν"; see G. S. Kirk, J. E. Raven, and M. Schofield, *The Presocratic Philosophers* (2nd ed.), Cambridge: University Press, 1983. The 'undifferentiated background' of phenomena can be identified with Anaximander's 'unlimited', άπειρον.

[2]   Plato, *Timaeus* 37d: "εικών κινητόν αιώνος"; see *The Collected Dialogues of Plato* (ed. by E. Hamilton and H. Cairns), Princeton: Princeton University Press, 1963.

[3]   Aristotle, *Physics* 219b: "αριθμός κινήσεως κατά το πρότερον και ύστερον"; see *The Complete Works of Aristotle*, Vol. I (ed. by J. Barnes), Princeton: Princeton University Press, 1984.

[4]   I. Newton, *Philosophiæ Naturalis Principia Mathematica*, Book I (transl. by A. Motte, rev. by F. Cajori), University of California Press, Berkeley, 1934. (Original work dated 1689.)

[5]   P. Galison in *Empires of Time. Einstein's Clock and Poincaré's Maps* (New York: Norton, 2003) links the late 19th century's development of transportation, communication and clock synchronization technologies with the conceptual development of Einstein's relativity.

[6]   An amusing simile between a clock and a gas-meter is found in the classical work of Victorian physics, *Elements of Natural Philosophy* by W. Thomson and P.G. Tait, § 367 (New York: Collier, 1902, p. 127).

[7]   *Cf.* A.S. Eddington, *The Nature of the Physical World*, Cambridge: University Press, 1928, p. 51.

[8]   E. Mach, *The Science of Mechanics. A Critical and Historical Account of Its Development*, LaSalle (IL): Open Court, 1960, p. 272 (Original work dated 1883).

[9]   Mach, ref. 8, p. 273.

[10]   H. Poincaré, *The Value of Science*, pp. 227–228; in *The Foundations of Science* (transl. by G.B. Halsted), New York: Science Press, 1929 (Original work dated 1898).

[11]   *Cf.* P. Mittelstaedt, *Der Zeitbegriff in der Physik*, Mannheim: BI-Wissenchaftsverlag, 1989, esp. Chapter 2.

[12]   E. Mach, *Analysis of Sensations* (transl. by C.M. Williams), New York: Dover, 1959, p. 349. (Original work dated 1886.) — Mach's proposal to measure durations as function of the test body's temperature requires that a temperature scale already has been defined; Mach himself studied the latter problem in his *Principles of the Theory of Heat*. Later we will see a construction of a time-keeping mechanism on a pre-metric basis.

[13]   J. Bergquist, S. Jefferts, and D. Wineland, *Physics Today* 54.3: 37, 2001.

[14]   See Bergquist *et al.*, ref. 13.

[15]   Incidentally, UT creates a great illusion of Newtonian time, 'flowing equably' in the background of all worldly processes, but not being related to any individual process in particular.

[16]   A. Carrel, *Science* 74: 618–621, 1931; P. L. du Noüy, *Biological Time*, New York: Macmillan, 1937.

[17]   B. Andresen, J.S. Shiner, and D.E. Uehlinger, *Proceedings of the National Academy of Sciences (USA)* 99: 5822–5824, 2002.

[18]   Taking exception to fundamentalist claims of 'ontological physicalism'; for a critical review see D. Papineau in *Physicalism and Its Discontents* (ed. by Loewer and Gillett), Cambridge: University Press, 2001.

[19]   T. O'Connor and H. Yu Wong, 'Emergent Properties', in *The Stanford Encyclopedia of Philosophy, Winter 2006 Edition*, (ed. by E.N. Zalta), URL = http://plato.stanford.edu /archives/win2006/ entries/properties-emergent/.

[20]   Speaking frankly: naïve reification of the notion of time in different domains creates false mysteries, and invokes vain attempts to solve them. The following quote provides a good example of such a 'mysterianism': "Both psychological and physiological times flow in the same direction. But their reciprocal relations remain as mysterious as those of consciousness and cerebrum" (Carrel, ref. 16, p. 621).

[21]   The case of spatially separated observers comparing data from their private phenomenal fields is not considered.

[22]   This is true at least in most natural conditions. Under special experimental arrangements, subjects may experience uncertainty as to the temporal order of two perceived events, or the perceived order may be even reversed with regard to the objective succession. *Cf.* for example, G.B. Vicario, in *The Nature of Time: Geometry, Physics and Perception* (ed. by

R. Buccheri *et al.*) Dordrecht: Kluwer, 2003, pp. 53–75; also M.C. Morrone *et al.*, *Nature Neuroscience* 8: 950–954, 2005. However, these artificial, experimentally produced phenomena are not of essential importance for the questions of our interest.

[23]   Issues of this section are elaborated in more details in the article by J. Wackermann in *Mind and Matter*, 6.1: 9–50.

[24]   E. Mach, ref. 12, p. 349.

[25]   See, for example, J. Cohen, *Scientific American* 211: 116–124, 1964; H.E. Lehmann, *Annals of New York Academy of Sciences* 138: 798–821, 1967; R. Ornstein, *On the Experience of Time*, Harmondsworth: Penguin Books, 1969.

[26]   K. Gloy, *Zeit. Eine Morphologie*. Freiburg: Alber, 2006, p. 26 (transl. JW).

[27]   For a more detailed argument see ref. 23. However, this difficulty does not prevent some psychologists from taking the construct of 'speed of subjective time' for its face value and making it subject of enquiries or experimental studies: *cf.* A.D. Eisler, H. Eisler, and H. Montgomery, in *Fechner Day 97* (ed. by A. Preis and T. Hornowski), Poznań: International Society for Psychophysics, 1997, pp. 143–148.

[28]   See U. Ott, 'Time experience during mystical states,' in this volume. Alterations of temporal experience in psychotic, meditative, or mystical states have been traditionally of interest for phenomenologically or existentially oriented psychopathology and psychiatry (von Gebsattel, Strauss, Tellenbach, and other authors). Recently, these phenomena have attracted attention of physicists as well: *cf.* M. Saniga, in *Studies on the Structure of Time: From Physics to Psychopathology* (ed. by R. Buccheri, V. Di Gesù, and M. Saniga), Dordrecht: Kluwer, 2000, pp. 137–166.

[29]   Mach, *Mechanics*, ref. 8, p. 275.

[30]   Poincaré, *Value of Science*, ref. 10, p. 224.

[31]   Psychophysics was defined by its founder, G.Th. Fechner, as "the exact science of the functional relations of dependence among body and soul, more generally, between the corporeal and the mental, the physical and the psychological, world". (Fechner, *Elements of Psychophysics*, transl. by H.E. Adler, ed. by D.H. Howes and *E.G.* Boring, New York: Holt, 1966. Original work dated 1860.) In spite of the seemingly dualist diction of this quote, Fechner's intention was to overcome the psycho–physical dualism. For an account of Fechner's metaphysics and natural philosophy, see M. Heidelberger, *Nature from Within* (Pittsburgh: University of Pittsburgh Press, 2004), esp. Chapters 3 and 6.

[32]   J. Wackermann and J. Späti, *Acta Neurobiologiae Experimentalis* 66: 245–254, 2006.

[33]   The pair of physical durations (s_1, s_2) for which the intervals are perceived as 'shorter' or 'longer' with equal probability ½, is called the 'point of subjective equality' (PSE).

[34]   J. Wackermann and W. Ehm, in *Fechner Day 2007* (ed. by S. Mori, T. Miyaoka, and W.Wong, Tokyo: International Society for Psychophysics, 2007), pp. 515–520.

[35]   Therefore this effect has been named 'subjective shortening in memory'; see J.H. Wearden and A. Ferrara, *Quarterly Journal of Experimental Psychology* 46B, 163–186, 1993.

[36]   Reprinted from J. Wackermann, *Mind and Matter*, 6.1: 9–50; based on data reported by J. Wackermann, J. Späti and W. Ehm in *Fechner Day 2005* (ed. by J. S. Monahan *et al.*), Traverse City: International Society for Psychophysics, 2005, pp. 359–364.

[37]   J. Wackermann, W. Ehm, and J. Späti, in *Fechner Day 2003* (ed. by B. Berglund and E. Borg), Stockholm: International Society for Psychophysics, 2003, pp. 331–336; J. Wackermann and W. Ehm, *Journal of Theoretical Biology* 239: 482–493, 2006.

[38]   From Greek κλεψύδρα = water-clock, composed of κλέπτω = steal, and ὕδωρ = water. Klepsydræ are known in two variants, the *outflow* model, used to delimit a constant time interval, and the *inflow* model, measuring elapsed time as a function of amount of water accumulated in a container. These two principles are here combined in the abstract concept of inflow–outflow unit.

[39]   We say that the model operates on a 'pre-metric principle'. An illustrative example: to compare two weights, we may use a spring balance and compare two numeric data (metric operation); or we use a lever balance that shows directly, which of the two weights is heavier (pre-metric operation). Fundamental measurement devices are *naturally* of pre-metric character; *cf.* note 12.

[40]   J. Wackermann, in *Endophysics, Time, Quantum and the Subjective* (ed. by R. Buccheri, M. Saniga, and A. Elitzur, Singapore: World Scientific, 2005), pp. 203ff. There is accumulating neurophysiological evidence for the proposed interpretation: *cf.* R. Jech *et al.*, *NeuroReport* 16, 1467–1471, 2005; O.V. Sysoeva *et al.*, *PLoS One* 5: e12650, 2010; O.V. Sysoeva *et al.*, *Frontiers in Integrative Neuroscience* 5: e37, 2011.

[41]   E.G., Boring, *The Physical Dimensions of Consciousness*. New York: Century, 1933, p. 136.

[42]   J. Wackermann, *Spanish Journal of Psychology* 10: 20–32, 2007.

[43]   Particularly the property of 'serial additivity' (J. Wackermann, *Journal of Mathematical Psychology* 50: 495–500, 2006), from which the qualities demonstrated in this section directly follow.

[44]   If a divided ruler is applied, the length can be 'immediately read out'; but the ruler is nothing else than a system of copies of the unit, fixed on a permanent material carrier.

[45]   The choice is only a matter of convenience. In the course of our analysis, only relations between readings of a given clocks ensemble are of interest, while the respective UT-coordinates *t* of the readings can be eliminated.

[46]   In the common understanding, a uniform scale is that which 'consists of equal parts'. In case of space measurements, this can be tested directly by comparing two shifted copies of a given scale. With time measurements, no direct test of equality of the clock's period is possible (*cf.* note 30). Comparisons of readings between 'shifted clocks'—*i.e.*, clocks started at different epochs—are possible and play an important rôle in the theory given below, but we will see that their interpretations are necessarily ambiguous.

[47]   J. Wackermann, in *Fechner Day 2005* (ed. by J.S. Monahan *et al.*), Traverse City: International Society for Psychophysics, 2005, pp. 353–358.

[48]   Note that for $\kappa \to 0$, eq. (2) becomes a linear function of s, $krf(s,w) = \eta s$, and with $\eta = 1$ the KRF becomes the identity function. In this limiting case the KC is a regular clock counting identical periods: $(\forall n)\ r_n = u$. Then trivially $t_n = nu$, $\vartheta(t) = t/u$, and eqs. (4,5) are obvious.

[49]   This is to read as $t_{B'} - t_B = a = t_{A'} - t_A$.

[50]   Reprinted from J. Wackermann in *Biological clocks* (ed. by O. Salvenmoser and B. Meklau), Nova Science Publishers, 2010, pp. 177–190.

[51]   Observe that the chronometric ratios λ and intercepts μ generally depend on the clocks' units as well as on their epochs. Only in the limiting case $\kappa = 0$ the initial conditions (epochs) separate from the instrumental constants (units); *cf.* note 48.

[52]   Assuming that the loss rate κ is approximately equal for all subjects.

[53]   This notion of objectivity as idealization-supported intersubjectivity makes our approach close to so-called 'protophysics'; *cf.*, for example, P. Lorenzen's essay 'How is objectivity

of physics possible?' in Lorenzen, *Methodisches Denken*, Frankfurt: Suhrkamp, 1968, p. 142–151. As for relations between protophysics and fundamental physics, *cf.* also Mittelstaedt, *Zeitbegriff*, ref. 11, p. 144ff.

[54] Originally published by E.A. Milne and G.J. Whitrow in the *Zeitschrift für Astrophysik* 10: 263–298, 1938, and later summarized in Chapter 2 of Milne's *Kinematic Relativity*, Oxford: University Press, 1948. For a non-mathematical exposition of the principal ideas see also Milne's *Modern Cosmology and the Christian Idea of God*, Oxford: Clarendon Press, 1952.

[55] Milne and Whitrow, ref. 54, p. 265.

[56] Milne and Whitrow, ref. 54, p. 266.

[57] Milne, *Astrophysical Journal* 91: 158, 1940. For example, in the expanding universe, on the t-scale, spectral red-shifts in the light arriving from distant galaxies are explained by the Doppler effect; in the stationary description on the τ-scale, they are interpreted by secular variation of Planck's constant h. See *Kinematic Relativity*, pp. 112–126; *Modern Cosmology*, pp. 134–145; *cf.* also G.J. Whitrow, *Monthly Notes of the Royal Astronomical Society* 114: 180–190, 1954.

[58] Milne, *Kinematic Relativity*, ref. 54, p. 50.

[59] G.J. Whitrow, *Observatory* 116: 265, 1996.

[60] Poincaré nicely commented on Carlyle's adoration for 'historical facts': "That is the language of the historian. The physicist would say rather: 'John Lackland passed by here; that makes no difference to me, for he never will pass this way again.' " H. Poincaré, *Science and Hypothesis*, p. 128; in *The Foundations of Science* (transl. by G.B. Halsted), Science Press, New York, 1929 (Original work dated 1902).

[61] This is what Mach so often emphasized; hence his idea of the ultimate unity between the law and the content of the universe; *cf. Mechanics*, ref. 8, p. 284.

[62] Milne, *Modern Cosmology*, ref. 54.

[63] That Milne's cosmology is a theory of communicable universe remained largely unnoticed by his commentators; a honorable exception is M. Johnson's *Time, Knowledge, and the Nebulae*, New York: Dover, 1947; *cf.* p. 43ff.

[64] Milne, *Modern Cosmology*, ref. 54, p. 160.

CHAPTER 9

God, Time and Eternal Life

Dirk Evers[*]

Martin-Luther-University, Halle-Wittenberg, Germany

Abstract: The essay intends to clarify the ontological status and validity of time as such, as well as the modes of time (past, present, future) and to relate them to God's eternity. The famous analysis of time and time-experience given by Augustine serves as a starting point to demonstrate the questions at stake for a theological reflection on the relation between time, human existence and eternity. The author then argues for the ontological significance of the passage of time and thus for the eternal significance of transient human existence by referring to physical cosmology and the philosophical concept of presentism. System theory and semiotics allow for further clarification of the status of temporal processes for the development of human existence and self-understanding. In that context, religion is seen as a means not to step out of time, but to form and devise human temporal existence for individuals and communities. Finally, the Christian triad of faith, hope and love is mapped to the Augustinian categories of memory, expectation and attention, thus providing the hermeneutical key for an eschatological perspective of human existence.

Keywords: Eternity, GOD, Augustine, presentism, eternalism, block universe, eternal life, J.E. McTaggart, relativity, Einstein, cosmology, thermodynamics, Peirce, semiotics, religion.

1. INTRODUCTION

Any theology that examines the Christian experience and formation of time has to give an account of God's relation to time. Human beings like all other forms of living things are temporal and transient beings. We only become what we are in and throughout time, just as we finally decay and die in and through temporal processes. Our existence is a finite existence between birth and death. However, God is eternal. God is not born, nor can God die. God is always what God is.

---

*Address correspondence to **Dirk Evers:** Institute of Systematic Theology, Practical Theology and Religious Studies, Martin Luther Universitäat Halle-Wittenberg, Halle (Saale), Germany; Tel: +49 34555 27088; Fax: +49 34555 23012; E-mail: dirk.evers@theologie.uni-halle.de

Some even claim that God is changeless. With that in mind, how could the eternal God relate to our temporal existence? How is God involved in the processes of creation and their contingencies? And how can we as finite, transient beings participate in God's eternity? In order to answer these questions we have to clarify the ontological status and validity of time as such, as well as the modes of time (past, present, future) and relate them to God's eternity. For that purpose, I will begin with a short reminder of the famous meditation on time which can be found in St. Augustine's Confessions (2.). These reflections will introduce important notions for our questions, while the following critical remarks will point to some fundamental flaws of the Augustinian analysis of the relation between time and eternity (3.). In the third and main part of my paper, I will attempt to tackle these flaws by arguing for the ontological validity of time (4.) and by reconstructing modes of time formation (5.). Building upon these results, the final paragraph will establish a constructive theory of the relation between time and eternity, which makes use of the notion of remembrance (6.).

2. AUGUSTINE: ETERNITY AS CONTRAST TO TIME

Augustine points to the apparent contradiction that we measure time while it passes, and speaks of time in terms of intervals like minutes, hours, days *etc.*, as if time were something with an extension. But simultaneously we experience time as impossible to hold or capture. The past has come and is no more, the future still remains to come and is not yet, and the present seems to be only the dimensionless borderline between past and future. If time consisted only of the present without a past, into which it disappears, or future that it realizes, the present would not be time, but eternity. Time "is" only, while it disappears and the future that isn't yet becomes the past that is no more. Thus it tends toward non-being [1].

Still, we measure time and speak of longer and shorter intervals of time. We experience time as the basic form of our existence as created beings, and we experience time not as an unextended transition, but as longer or shorter. This is possible only because time "passes by" [2]. We cannot measure past or future as past or future: we have direct access only to the present. And the past is present to us only insofar as it has left traces in our memory, while the future can only be present as mental anticipation of things to come. From these observations Augustine derives his answer to the question of what we measure when we measure intervals of time. Time, he says, is the extension of the mind (*distentio animi*) [3]. It is in the human mind where the three modes of time come together in such a way that time becomes a fabric, in which we can measure the length of time: "It is in you, Oh mind of mine, that I measure the periods of time … I measure as time present the impression that things make on you as they pass by and what remains after they have passed by—I do not measure the things themselves which have passed by and left their impression on you. This is what I measure when I measure periods of time" [4]. The human mind consists of three faculties: memory, attention and expectation. It remembers past events, pays attention to present events and expects or anticipates future events. And in doing so the human mind captures the passing of time and transforms it into duration, into extended forms of perception. The most important phenomena through which we can verify this analysis are music, rhythm and the understanding of language. While singing a song or reciting a poem, the human mind somehow comprises the whole of what has already been sung or said in memory and what will be sung or said in expectation and, at the same time, is fully present by giving sound to one note or word after the other. Thus, the human mind is able to condense the constant flow of impressions into meaningful entities.

For Augustine, however, the human experience of time is also the experience of contingency and disruption. Because human existence is existence in time through memory, attention and expectation, it is distorted and torn in itself between past and future. The distension of time which we experience in our minds is one of the essential features of our alienated existence far from God. It is so fundamental for human life that Augustine summarizes: "Behold, my life is but distension" [5]. In this passage Augustine refers to the fact that distraction is the second lexical meaning of the Latin word *distentio*. As temporal beings we lose ourselves to time, which overstretches and overstrains us and distracts us from God's eternity. To cite Augustine again: "I have been torn between the times, the order of which I do not know; and my thoughts, even the inmost parts of my soul, are mangled within tumultuous varieties" [6]. A concentration on the divine presence, an 'intention', must transcend the 'distension' of temporal existence so that we do not surrender to the distracting flow of time, but strive towards eternity: "non secundum distentionem, sed secundum intentionem" [7].

For Augustine, time is in sharp contrast to God's eternity. God's eternity is nothing but presence (totum esse praesens), a presence without past or future (nec futura nec praeterita) and therefore, is at a complete standstill (semper stans) [8]. To experience God and to overcome human distortions and disruptions means to be drawn out of the temporal flow of events into the timeless vision of God's eternity. For Augustine, time is not the mobile image of eternity as it is for Plato or Plotin. Time is not directed toward eternity, it is overcome by eternity. Human vision and audition are temporal senses that provide us with ever fluctuating and imperfect knowledge, while God as the eternal "creator of all times", as Augustine explains in De trinitate, has perfect and immediate knowledge of all times without this knowledge being temporal [9]. God's knowledge is non-transitional [10]. God knows everything, but as a perfect possession of knowledge and not in terms of memory and expectation.

However, Augustine cannot totally renounce the meaningfulness of time in its relation to God and at this point, a tension appears in his account of time [11]. Even though the human mind is only capable of gaining ever fluctuating and changing knowledge, it still builds up a persisting knowledge of self-identity within this flow of time. Therefore, it is again in the mind with its faculty of memory where the encounter between the temporal human being and the eternal God takes place. Augustine writes: "And thus since the time I learned of thee, thou hast dwelt in my memory, and it is there that I find thee whenever I call thee to remembrance, and delight in thee" [12]. In the inner sense (intima mea), the inner eye of the soul (oculus animae meae) can encounter eternity [13] and be lifted up out of space, time and matter through the love with which it adheres to God. And even when the soul then falls back into space, time and matter, it still carries God with it in its memory now being aware of its real home in eternity, beyond space and time [14].

3. "ONLY THROUGH TIME TIME IS CONQUERED": A MODERN CRITIQUE

Although Augustine analyzed the relationship between time, memory, consciousness and faith on a complex level still worth studying, his abrogation of time as meaningful as well as his sharp dichotomy between time and eternity are problematic for the modern mind. This is not only due to philosophical and scientific reasons, but also to theological ones. One can identify three main areas of concern, which I will refer to as the questions of *the actuality of time*, *the socio-biographical dimension of time* and *the relation between time and God's eternity*. I shall address these three aspects of a modern critique very briefly in order to deal with them more extensively and try to respond to them constructively in the third part of my paper.

3.1. The Actuality of Time

Modernity has discovered time as the main means of creation. The flow and dynamics of time have replaced the notion of a creator and have dissolved the idea of a master plan of creation. The discovery of the irreversibility of time in thermodynamics coincided with the discovery of natural evolution told as the story of how living organisms slowly developed over long periods of time without any teleological principle. The interplay of chance and necessity within time became a central notion for the scientific view on nature. Theories of self-organization in the second half of the twentieth century continued this trend and reconstructed ways in which structured systems and organized forms emerge through temporal processes. Einstein's theory of relativity merged time and space and pointed to the relativity of simultaneity. At the same time, Heidegger's existential philosophy of "Being and Time" focused philosophical thinking on time as the fundamental mode of existence. Time itself became a concept, which evolves and changes within time so much that some have even spoken of the temporalization of time [15]. All these developments point out to the fact that the actuality of time is an indispensable feature of the modern world view.

3.2. The Socio-Biographical Dimension of Time

These formations of modernity are accompanied by a change in the ways human beings integrate their life-span into the time of creation [16]. Lifetime and world time began to separate with the Copernican revolution after which it became obvious that the individual span of life is reduced to nothing given the enormous timescales on which cosmological, biological as well as cultural evolution act. While Descartes and others in early modernity still speculated upon the possible completion of science within a lifetime, the progress of science, of history, and of culture, changes in technology, in political systems and economics have proved that such a vision is an illusion. We are part of processes of long-time development, which are far beyond our individual scope of existence. Furthermore, we are facing possibilities, which we cannot exhaust within our individual lifespans. All of that has brought about an extreme acceleration of time and a constant urge to use time effectively, to develop time-saving methods and a preoccupation of the modern mind with the presence, with opportunities to grasp or miss out. Time has become something that is limitless and empty as such (cf. Kant's concept of time as the *a priori* form of the inner sense), but is a limited resource for the individual person and has to be filled, consumed and used by this particular person.

3.3. The Relation Between Time and God's Eternity

For Augustine, time is a main feature of creation and an indicator of its unsteadiness and transitoriness. It is opposed to God's eternity insofar as God is transcendent to time. Time means constant change and distension, while God is changeless, unmoving and unmoved. Time has a beginning and an end, while God is infinite and endless. Consequently, Augustine defined the aim and intention of human existence as the redemption from temporal existence, which is only fully realized beyond our existence in time where our restless hearts find peace in the eternal and unchanging union with God [17]. It can be anticipated in moments of beatified visions of God [18] when the human mind is taken out of time. Then time is superseded and dissolved into a purely intuitive perception of the eternal presence of God. All creation is nothing but a context of signs pointing toward God's eternal reality. The ultimate goal of human existence – its unification with God – is not reached within time, but only outside of time [19].

However, this is no longer plausible because of theological reasons and the modern awareness of the significance of time. Since Hegel, at the latest, the idea of the timelessness of God was more and more dismissed. The main argument was that God as a living God cannot be thought of as existing outside of time. The voice of Paul Tillich speaks for many: "If we call God a living God, we contend that he incorporates temporality and therefore, a relation to the modes of time"

[20]. If we do not want to think of God as an abstract idea, we have to work out a positive relation between God and time. That is also required by the biblical traditions which speak of God as a living God who creates the world, who interacts with his creatures, who reacts to human prayers and to human agency with joy, wrath, mercy, condemnation and redemption. God is the Lord of time and life and thus relates to time in many ways. God is affected by temporal developments, and according to the New Testament, God also becomes involved in time through incarnation. The eternal word became flesh, was born and died and thus, God himself participated in the temporality of human existence.

We can summarize these critical objections against a dichotomy between time and eternity to the effect that we overcome time with its unsteadiness and transitoriness not by abolishing time, but by using and designing time. Or to use the words of the poet Thomas Stearns Eliot: "Only through time time is conquered" [21]. At least in a Christian theological view faith and religion should not be seen as a means to step out of time, but to form and devise our temporal existence. Only time that has a shape (Gestalt) can enter into a constructive relation with eternity. And eternity, on the other hand, is neither the infinite prolongation of time nor the 'presence' of a timeless realm, but the power of conquering time in and through time.

To overcome the flaws of the traditional dichotomy between time and eternity, we have to deal with the three objections mentioned in such a way as to bring them into fruitful dialogue with theological notions. For that purpose we have to explore the ontological foundation of time within our modern scientific worldview in a theological perspective (4.). We then need to deal with the ways in which we shape and structure our time and modes of existence through faith (5.). And in the third and final step we will ask how God's eternity can be perceived in relation to our temporal existence so that the eschatological hope of faith becomes meaningful (6.).

4. TIME'S ONTOLOGICAL SIGNIFICANCE

For Augustine time had its ontological place only in the human mind (anima/animus). However, modern philosophy has shown the richness and interplay between time as a physical parameter and time as an experience of the flow of time. Both concepts of time are important and both are of ontological significance. Augustine's reference to the human mind is an important reminder of the fact that we know of time primarily through the temporal order of our own experience of time. "Indeed our understanding of the basic terms of the vocabulary of time is obtained from the temporal order of thoughts and perceptions" [22]. But from this fact it can not be inferred that the passage of time is nothing but a subjective illusion. On the contrary, reality must be conceptualized in such a way as to make the emergence of temporal experience possible.

In 20th century philosophy, this problem was at the centre of a debate on the tension between the dynamic experience of time and the apparent static parameter of time as it is used in empirical science. In a famous paper published in 1908, J.E. McTaggart argued that the experience of time in its succession of past, present, and future is a mere appearance [23]. He distinguished between two ways in which positions in time can be ordered, namely in the A-series and the B-series. The A-series of time, which represents our experience of time, is ordered according to the tenses of time and differentiates between *past*, *present* and *future* events. On the other hand, the B-series of time uses a two-place relation and orders two events in time according to the event x_1 being *before*, *after*, or *simultaneous with* another event x_2. McTaggart argues that there can be no reality of experienced time without an A-series, but that any A-series is inherently contradictory and therefore, represents no objective reality. Only the B-series has an ontological foundation and therefore, represents all that "time" is about. But "time", as represented by the B-series, is static time because it only refers to positions on a time scale and does not presuppose an objective passage of time.

Many philosophers have accepted McTaggart's distinction between the A- and the B-series of time, but not all of them accept his rejection of the A-series. Those who follow McTaggart in rejecting the reality of a passage of time and instead conceptualize time only as an ordered series (B-theorists), reduce absolute properties about past, present and future to relational properties of before, simultaneous and after. For example, when we ascribe the property of being past to the 15th century, this can only be understood in the sense that the 15th century is earlier than the 16th, the 17th, and the time in which we are speaking, the 21st century. But it is not "past" as such, because the passage of time is an illusion and time has to be understood as fundamentally 'tenseless'. For the B-theorist time is static and conceived in analogy to a spatial dimension. Just as space is the order of things insofar as they exist simultaneously, time is the order of things insofar as they exist successively [24]. And just as there are no absolute spatial properties like "being north", but only relational properties like "being north of", there are no genuine tensed properties of time, but only relations such as before and after.

The opponents of the B-theory would normally concede that there can be no reality of time without an A-series of time, but they would reject the claim that the A-series is contradictory, a position that is often referred to as the A-theory. For an A-theorist, the tenses of time are objective and founded in reality so that knowing all facts about before, after or simultaneity would not give full information about what time is and about how past, present and future are alike. While the static 'tenseless' view of the B-theorist takes past, present and future to be only psychological phenomena of our perception of time, the dynamic view of the A-theorist insists that this perception has a *fundamentum in re*, which is grounded in a distinction in reality.

Yet, this theory can come in different versions [25] regarding the ontological foundation of time, so that even some A-theorists hold that the experience for the flow of time is an illusion. Those A-theorists have argued that we cannot make a meaningful ontological distinction between past, present and future entities. Not only when we refer to present objects and events, but also when we refer to factual events of the past and even to future possible events we have to presuppose that our references have an objective truth-value. We therefore have to grant future and past objects and events an ontological status on par with the present. For those A-theorists space-time exists as a complete four-dimensional block universe which is one given single entity. Consequently, the appearance of a passage of time, of things coming to be and ceasing to be, is only due to the movement of our consciousness in time and space so that "the present is like a spotlight moving across the whole block of time" [26]. This view is often referred to as *eternalism* or the *moving spotlight view* [27].

Representatives of the moving spotlight view are often found among physicists who conceptualize reality along the lines of Einstein's theory of relativity. The theory of special relativity intermingles space and time in a four-dimensional space-time, or with the famous words of Hermann Minkowski, who gave relativistic space-time its mathematical form: "Henceforth space by itself and time by itself are doomed to fade away into mere shadows, and only a kind of union of the two will preserve an independent reality" [28]. The theory of relativity dissolves a strong notion of simultaneity so that it seems difficult, if not impossible, to ascribe a privileged ontological status to 'present' objects or events. What is simultaneous with other events is relative to the observer's reference frame, which in special relativity depends on its relative velocity. Thus, it becomes impossible to identify spatial planes of simultaneity that are fully disentangled with time.

Einstein and others took this as an indication of the illusionary character of dynamic becoming. The notion of spatial relations changing within an objective passage of time seemed untenable for Einstein: "Physics changes from *becoming* in the three-dimensional space to *being* of the four-dimensional 'world'" [29]. The four-dimensional block universe is the static whole of reality, and only for the human mind, which moves inertially at a speed much less than light, the four-dimensional continuum is separated into three spatial dimensions, which change in time. The great mathematician and interpreter of Einstein's theory of relativity, the German mathematician Hermann Weyl, once formulated: "The objective world doesn't happen; it plainly is. Only for the view of a consciousness, which creeps along the world line of its body, a sector of this world becomes 'alive' and passes along as a spatial image which changes in time" [30].

But there are a number of arguments that call these conclusions of the illusionary nature of time experience into question and argue for the ontological validity of both the A-series and the B-series of time. We can not deal with them extensively and have to concentrate on some important issues. First of all, we note that the very design of relativistic space-time reveals its highly constructive and preparatory character. In order to merge space and time into relativistic space-time, the physical parameter of time must be transformed beforehand into a spatial one so that its integration with the three spatial dimensions into the four-dimensional space-time becomes possible [31]. Therefore, the design of the theory of relativity is such that time is transformed into a spatial category and thus, deprived of its tenses from the very beginning. It is an unjustifiable, subsequent reification or hypostatization to understand the geometrical Minkowski-world as a representation of a given block universe of world-points and to infer ontological claims from its geometrical structure.

The geometry of space-time in the theory of relativity indeed points to McTaggart's B-series of time only, but if the boundary conditions provided by the principle of the constancy of the speed of light is taken into consideration there is an objective network of causal relations so that the succession of cause and effect of two linked events is the same for every observer. A reversal of cause and effect between two events in time-like distance is excluded. The objective aspect of the relativistic notion of time is the order of the causal network [32]. There is a universal temporal order at each moment in time and time in this sense, as the order of cause and effect, is one-dimensional.

Furthermore, the extension of the theory of special relativity to the theory of general relativity has revealed the intrinsic relatedness between space-time and material and energetic objects and events. In general relativity, distances in space and time have become dynamic physical parameters which are closely interlinked to the physical energetic processes. Real space-time is not like the block universe, *i.e.,* the receptacle or container notion of space-time, which provides the geometry and metric form into which otherwise relationless objects and events are introduced. Physical objects and events do not inhabit space and time like an apartment. As early as 1854 the German mathematician Bernhard Riemann, whose geometrical work provided the basis for Einstein's general theory of relativity, voted against the notion, "that the metric of space is fixed independently from the physical processes happening in it and that reality would move into this metric space as into a tenement" [33]. But then, space-time is deeply related to the process of events and thus participates in it.

However, the claim that the experience of a passage of time is an illusion and that past, present and future objects and events have the same ontological status inevitably leads to the conclusion that the future is fully determined and that time does not open up new possibilities. The difference between the real world and accessible possible worlds gets blurred and intractable. This is in sharp contrast to the progressive developments we see in modern cosmology as well as in theory of evolution. Any development in time that enters new realms of structure and form can not be reduced to a simple order between earlier and later, but must be interpreted in terms of an objective past, which provides the conditions for present possibilities, which in turn, provide accessibility to contingent future possibilities [34]. A reduction of time to a subjective experience in a block-universe would deprive cosmology, evolution as well as human agency – all directed from past towards future – of their meaning, including the striving towards scientific progress itself.

Other A-theorists, however, have argued that eternalism cannot account for the openness of the future with its undetermined possibilities. While our references to past facts and present events are the basis for determined truth-values, this does not apply to what will be the case in the future. Consequently, past as well as present objects and events have to be regarded as real, while future events are still unreal and yet to come. Those A-theorists endorse what has been called a *growing block* view of time or the *growing universe theory* [35] in which the past and the present are real, but the future is only possible. In this view the flux of time is the growing of reality at the edge of the present where the future gets realized and thus adds to reality. Though the future is not real, in this view past and present still share the same ontological status. Again the present of our experience of time is very different from the objective reality of past objects and events, and the inaccessibility of the past for our direct experience remains a mystery in this view.

In the light of our experience of the flow of time, of our theories of physical development in time and of theological considerations I opt for the third interpretation of the A-series which is called *presentism*. *Presentists* are convinced that there is a fundamental ontological difference between past, present and future insofar as only the present exists in an emphatic sense. The past once existed, but does not exist anymore, while the future does not exist yet. Socrates *e.g.,* does not exist any more because he is dead and not temporally present, while future generations such as my great grandchildren, do not exist yet (if they ever come into existence at all). This does not imply that we cannot assert truth definite sentences with regard to past or future object and events, but only that the corroboration of such truth claims is different in each case. When we refer to contingent facts of the past, we claim that this statement can be verified by means of inference from present facts, *e.g.,* from documents that have come to us. When we refer to contingent possibilities in the future, we can only claim a certain probability for our prediction. A final verification is only possible when what has been predicted becomes accessible through experience or facts. The ontological difference between the present and the other modes of time lies in their different accessibility. To present objects and events we stand in a relation of direct access, *i.e.,* of directly experiencing them as they are. We perceive a tree as this tree only if it is present, or in other words, to perceive something as present is to perceive it. To past things and events we can only relate to in terms of indirect experience, *i.e.,* in ways of remembrance, of historical reconstruction, of inference from present data to past condition that account for them *etc.* And we can make statements about future objects and events only through extrapolation from present and past experience and thus by ways of more or less reliable predictions. Any realistic theory of time which takes our experience of time and transient existence seriously must be able to integrate our modal notions of actuality, necessity and possibility [36].

And finally, only the interplay between the A-series and B-series of time can give an account of the existential dimension of human existence. Human beings are able to relate to time and to deliberately orientate themselves in time. As conscious and subjective beings we exist in time (B-series) and relate to time (A-series) insofar as we relate to things which are past and therefore not directly accessible, or presently at our disposal or which are yet to come. We are not passive sufferers of time; as self-reflective beings we are present in time and that includes the possibility to bring about things and events in time. Our existence is such that we can shape and form the passage of time. How can we further understand this fundamental feature of our reality?

5. SHAPES AND FORMS OF TIME

Our standard cosmology of an expanding universe refers to an 'age' of our cosmos and presupposes a cosmic foundation of a time-space relationship. I abstain from discussing the different models with regard to big bang theory such as cosmic inflation or the question of the geometry of cosmic space-time and the ratio between baryonic matter, cold dark matter and dark energy *etc.* I concentrate on the fact that an expanding universe is presupposed in all possible models. The expansion of the universe establishes and designates a cosmic time and prevents the cosmos from collapsing and from reaching an overall thermodynamic equilibrium too soon. "Any gravitating universe that can exist and contains more than one type of interacting material, must be asymmetric in time, both, globally in its motion, and locally in its thermodynamics" [37]. Therefore the theory of relativity in connection with cosmology promotes an understanding of space and time in relational terms. The universe is a relational process in itself which continually transcends its states and makes new possibilities accessible.

Thermodynamics teaches us that all complex, organized systems are sustained through dissipative processes, and that in the series of these processes, bifurcations occur, which lead to the emergence of more and more complex systems. The identity of organized systems, which are the result of constant energy and matter flows, is not an identity of substance, but an identity of structure, which is built up and sustained in time, then consummated in time as well. Only through temporal processes identities of organisms develop and persist in time.

Broken symmetries, feedback loops of causality and top down constraints bring self-organizing systems into being and provide the foundation for different scales of time. Ilya Prigogine has rightly argued that the theory of dissipative structures "permits us to distinguish various levels of time: time as associated with classical or quantum mechanics, time associated with irreversibility ..., and time associated with 'history' ... I believe that this diversification of the concept of time permits a better integration of theoretical physics and chemistry with disciplines dealing with other aspects of nature" [38].

Non-linear systems evolve in the irreversible and historical process of the developing cosmos, and in the case of organisms and self-conscious beings, the causal chains get entangled in such a way that they establish over-complex knots within this net. The history of evolution on our planet demonstrates how organisms are characterized by their ability to partly disconnect themselves from their environment, sustain their own form and thus become able to relate to environmental changes and challenges. Organisms are not simple functions of external parameters. They can outlive temporal changes and protect themselves, plan for themselves and make provision with respect to future needs.

This development reaches a new level in humans as subjective and self-reflective beings. The human mind evolves through a complex process of phylogenetic and ontogenetic developments and gives rise to a notion of self and responsibility. A necessary and essential condition for this process seems to be language and the interaction between human beings, through which we ascribe agency to others and to ourselves, so that we begin to say: That was *me*. *I* did it. *I* didn't want *you* to do it. In this process, we learn to articulate ourselves, to refer to ourselves as distinct from the circumstances, and ask for motives and purposes and develop criteria with which we start to judge our decisions and shape them accordingly. This provides the basis for human beings in their relation to time.

Thus, all entities that come into being through the interrelatedness of the network woven by time participate in the semiotic dimension of reality. Events and systems can refer to each other; they can represent and relate to each other in ways, which finally open up the space of consciousness. Charles Sanders Peirce has analyzed the basic formation of the semiotic sphere in form of a triad of categories, which he calls *Firstness*, *Secondness* and *Thirdness* [39]. Firstness (quality) represents existence as such without relation to others; Secondness (relation) comes into being when the changes of one entity depend on the changes of another; Thirdness (representation) then allows for all full-fledged semiotic processes in which something changes in dependence to another entity or event while this interdependent change is represented for a third party. Thus, self-reference becomes possible *via* the coupling of processes which become signs for others.

As Augustine has shown, human beings as self-identical beings relate to time through memory, attention and expectation. In contrast to material objects as well as plants and animals, human beings do not only exist in time, but relate to time as well. They use time as a tool and means of orientation; they are not only taken along with time. The foundation for human time management is laid through the basic faculties of perception and cognition, then it is fostered and processed through language and it culminates in building our autobiographical selves embedded in and contrasted with the narratives and concepts of tradition.

As human beings we participate in the evolution of time in a specific way, in the mode of self-referential beings. We do not only design our individual and social systems in time, we design time itself insofar as we handle time like a resource. Among all creatures only human beings can 'find' time, 'take' their time or 'lose' time. They even need leisure time or time off in order to have time for themselves, to find themselves in time. This ability to relate to time and to deal with time as a means of structuring and forming our existence is something human beings have to learn, and the process in which they learn it is the awakening of the human mind.

But, our scientific concepts of time participate in this structure as well. Empirical science as the interplay between conjectures and refutations is always inferring predictions of future events from an analysis of past events. Although science aims at establishing invariant laws which then describe the respective dynamics in time, it is in itself nothing but an endeavor in time to sketch mathematical models of reality which are only valid for the time being and subject to change and enhancement any time. Our scientific activity is a means of orientation in time and thus, is as an activity not reducible to the B-series as McTaggart suggested.

Thus time is a multifaceted construal. What we call 'time' in everyday language "is not a simple or primitive concept, but is rather a construct, which derives from several different sources" [40]. Therefore, we have to distinguish between different concepts of time and have to be aware of the fact that our temporal existence takes place in and through the interplay of these different concepts. That is especially important for our notion of presence as the basic category in which we experience the flow of time. It does not designate a specific temporal moment in time such as the indexical term 'now'. Presence in its full sense means the presence of something for someone. But this presence is neither the property of the thing present nor the property of the one for whom it is present. It is a relational concept and depends on the interrelatedness of events in space and time.

It comprises aspects of accessibility, of past conditions for and future consequences of something being present to someone. It can not be reduced at any rate to the concept of simultaneity.

Presence is not only a relational category; it is also a fundamental concept for any human community. We coordinate our existence in the presence of others *via* social and cultural time management. We relate different natural, private and public events according to formal systems such as calendars, clocks and timetables. And we not only coordinate our presence with the presence of others so that we can act accordingly, we need these relations to direct our agency from past, factual conditions to future possibilities and opportunities. It is at this point where religion comes in.

One significant feature of religion is its influence on the shaping and organization of time. From time immemorial religion has shaped human temporal existence and provided orientation by differentiating between relevant and irrelevant distinctions in time. Religious beliefs form individual and communal patterns of time by establishing festive seasons, recurrent structures of lifetime like weeks, months *etc.*, which provide time for work, rest and worship, and the ritualization of temporal ceremonies like services and prayers and thus have a strong impact on the ways their followers shape their individual lifetimes. In the Christian tradition the Sunday, the ecclesiastical year with its alternation of fasting and feast and the schedules of services with prayer, singing and preaching have had a tremendous effect on Christian culture. Biological and seasonal rhythms, basic physical features of time and anthropological essentials are linked within a theological framework to God's own relation to time. God's creational acts, his agency in history, his presence for the believer, and the expectations of his future deeds are all points of reference for the ways in which Christians as individuals as well as Christian communities shape their temporal existence.

Relating to time implies certain degrees of freedom. Through our relation to time we distance ourselves from time and gain a relative independence of external determination, which opens a space for spontaneity. Our agency is not simply a function of time determined by fixed preconditions. Faith serves the neighbor "freely and spontaneously" [41]. Deeds of love and compassion know to wait for their time (καιρός); they are not triggered like an acquired reflex, but grow out of the individual personality which is cultivated *via* memory, attention, and expectations.

This is evident when we become aware of the fact that we as human beings are able to interrupt our functional routines – an ability, which is inappropriate in many contexts of technology. Human existence is not the execution of a programme in time. Human beings can stop themselves when they find themselves functioning in an alienated and dull routine. We get lost when we simply deliver ourselves to the flow of time. In time we step out of time by reflecting reasons, means and ends.

It is an important function of religion to ritualize such kinds of interruption of our everyday routines through services, prayer, meditation, listening and singing. Religion is a means of structuring time through interruption and repetition [42]. And any religion that is able to develop and foster relative freedom provides means of intermediation between rules and spontaneity, between law and gospel, nature and grace, imperative and indicative, tradition and renewal by reflecting time in its relation to eternity.

6. TIME AND ETERNITY

In a theological reflection on time that takes these insights seriously it is important to define the relation between time and God's eternity accordingly and in such a way that God's presence is shown as God's participation in our formation of time [43]. God's eternity is neither the negation of time nor its suspension. It is not mere timeless, but it gets involved with time. Time can be conquered through time because eternity participates in time. Paul Tillich made the point that "the Divine Being is not a being beside others. It is the power of being conquering non-being. It is eternity conquering temporality. It is grace conquering sin. It is ultimate reality conquering doubt" [44]. God's eternity can be defined as the power with which God is present and active in creation in order to establish formations within time that are meant to participate in God's eternal being.

In a Christian perspective, God's relation to time is triune. It encompasses all three modes of time, and it does so through creation, transformation and perfection. The eternal God provides time for creation, transforms creation through time and employs time for his ends. The Christian faith responds to the eternal God in time. It identifies the creator as God the Father, whose faithfulness accounts for the reliability of the temporal processes of creation, which we cannot produce, but to which we owe our existence. Another aspect of faith is the trust in the transformational power of love. Amidst the transitory time of existence, forgiveness and renewal are possible. In these transforming events, faith identifies the salvific activity of God the Son, the redeemer, who discloses the meaning of existence and transforms us into the image of God. In the openness of the future and the self-transcendence of time toward fulfillment, faith identifies the perfecting activity of the Spirit who wins us and guides us toward the ends of God. That way God is present in time with his eternity while at the same time the process of creation enriches God's eternity.

Therefore, Paul's famous triad of faith, hope and love in 1 Corinthians 13 can be seen as a summary of the believer's relation to time: *faith* as the commemorative trust and belief in what God, from whom all things come, has provided in creation; *hope* as the longing anticipation of what God is going to bring about, historically and eternally; *love* as the attentive and transforming care for our fellow human beings, for ourselves, for the whole creation of which we are part. The eternal God is present to us and is intermingled into our existence by being mindful of us, by caring for us and by luring us into the future.

Through the complex interplay of faith, hope and love, Christians form and structure their existence in time. As individuals and as communities, they are enabled to cope with times of distraction, of doubt and perplexity. Faith, hope and love can bring determination to unassertive situations through memory and expectation and thus, can focus on present intentions. The believers can develop new perspectives of action, confidence and trust. They can experience that through *memory*, *expectation* and *attention* God himself gets involved in human history and cooperates with them. With commemorative, attentive and expectant love human beings cooperate with the triune God and in doing so, they actualize their destination as images of God. That is not contradicted by the finitude of human existence; rather we can recognize in our limited lifetime with birth, life and death, a distinctiveness and concreteness which is eternally meaningful in the eyes of God.

If this concept of eternity conquering time through time is adequate, the eschatological hope for our individual lives need not be grounded in the immortality of a substance-like soul. It is not that the material human body decays while an immaterial eternal soul remains. Eternal life is not an additional, never-ending life-span after death. It is the 'essentialization' [45] of our lived lives into a new form of everlasting existence that participates in the eternal life of the triune God. Even when in the process of aging and dying, our soul and mind disintegrate in the same way as our body, God will be mindful of us because his eternal life participates in our transient existence. God recalls what we strived for, where we failed, what – in good terms and in bad, in richness and deprivation – has shaped and formed our concrete individual existence. God as creator, redeemer and sanctifier will judge, heal, transform and concentrate our lived lives into a new form of existence so that we will be able to enjoy the everlasting presence of God because then time is finally conquered.

ACKNOWLEDGEMENTS

Declared none.

CONFLICT OF INTEREST

The author(s) confirm that this chapter content has no conflict of interest.

REFERENCES

[1]   Augustine, Confessions XI, 14, 17: "tendit non esse: tending no to be".
[2]   Conf. XI, 16, 21: "praetereuntia: as they are passing".
[3]   Cf. Conf. XI, 26, 33: "Inde mihi visum est nihil esse aliud tempus quam distentionem: sed cuius rei, nescio, et mirum, si non ipsius animi: Whence it seemed to me, that time is nothing else than distention; but of what, I know not; and I wonder, if it be not of the mind itself." And a little later Augustine concludes: "In te, anime meus, tempora metior: it is in you, O mind of mine, that I measure the periods of times" (Conf. XI, 27, 36).
[4]   Conf. XI, 27,36: "In te, anime meus, tempora metior. … affectionem, quam res praetereuntes in te faciunt et, cum illae praeterierint, manet, ipsam metior praesentem, non ea quae praeterierunt, ut fieret; ipsam metior, cum tempora metior."
[5]   Conf. XI,29,39: "ecce distentio est vita mea".

[6]   Conf. XI,29,39: "ego in tempora dissilui, quorum ordinem nescio, et tumultuosis varietatibus dilaniantur cogitationes meae, intima viscera animae meae".

[7]   Conf. XI,28,37.

[8]   Conf. XI,11,13.

[9]   Augustine, De trinitate II,5,9: "operator omnium temporum. ordo … temporum in aeterna dei sapientia sine tempore est."

[10]   Conf. XII,15,18.

[11]   Cf: See below endnote [19].

[12]   Conf. X,24,35 (following the translation of Outler, A.C. (1955). *The Confessions of St. Augustine*. Philadelphia: Westminster Press).

[13]   Cf. Conf. VII,10,16.

[14]   Cf. Conf. VII,17,23: "Sed mecum erat memoria tui … inveneram incommutabilem et veram veritatis aeternitatem supra mentem meam commutabilem: But thy memory dwelt with me … I realized that I had found the unchangeable and true eternity of truth above my changeable mind."

[15]   Cf. Sandbothe, M. (1998). *Die Verzeitlichung der Zeit*. Darmstadt: Wissenschaftliche Buchgesellschaft.

[16]   Cf. Blumenberg, H. (2001). *Lebenszeit und Weltzeit*. Frankfurt a.M.: suhrkamp.

[17]   Cf. the famous passage at the beginning of the Confessions, Conf. I, 1, 1: "inquietum est cor nostrum, donec resquiescat in te". The *donec* must not be understood in a consecutive sense, cf. Schmidt, E.A. (1985). *Zeit und Geschichte bei Augustin*. Heidelberg: Carl Winter Universitätsverlag, p. 42f.

[18]   Cf. Conf. IX, 10, 25, where Augustine speaks of the momentum intellectum. See also Chapter 7 in this eBook.

[19]   It has often been remarked that at this point a tension occurs between Augustine's notion of time in the meditation of the Confessions and the notion of time in De Civitate Dei. For our purpose we can refrain from further discussing this matter.

[20]   Tillich, P. (1956). *Systematische Theologie vol. 1 (3rd ed.)*. Berlin/New York: de Gruyter, p. 315.

[21]   Eliot, T.S. (1971). *Four Quartets*. Orlanda Fla.: Harcourt, p. 16 ("Burnt Norton", line 89).

[22]   Denbigh, K.G. (1981). *Three Concepts of Time*. Berlin/Heidelberg/New York: Springer, p. 166f.

[23]   Cf. McTaggart, J.E. (1908). The Unreality of Time. *Mind, 17*, 457–474.

[24]   Cf. Leibniz, who was the first to understand space and time as purely relative categories: "… je tenois *l'Espace* pour quelque chose de purement relatif, comme *le Temps*; pour une ordre des Coexistences, comme le temps est une ordre de successions: I regard space as something purely relative, like time; as an order of coexistence, just as time is an order of succession" (Leibniz, G.W. (1978). *Die philosophischen Schriften vol. VII*. C. I. Gerhardt, (Ed.). Hildesheim: Olms, p. 363; Third Letter to Samuel Clarke).

[25]   Cf. Peirce, Ch. S. (1992). *Reasoning and the Logic of Things. The Cambridge Conferences Lectures of 1898* K.L. Ketner (Ed.). Cambridge, Mass.: Harvard University Press. See also Dalferth, I.U. (2006). *Becoming Present. An Inquiry into the Christian Sense of the Presence of God*. Leuven/Paris/Dudley, MA: Peeters, p. 53f.

[26]   Pierce, J. (2003). Review of God and Time: Four Views. *Faith and Philosophy, 20*, 504–509, p. 505.

[27] One can also distinguish two versions of the moving spotlight view, one in which past, present and future events are fully fixed so that there is only one real world. Another in which the universe is not actually a complete four-dimensional block but a whole set of possible worlds which are all linked with each other. The spotlight of our present moves across the branching possible worlds and finds its path through branching possibilities. In this view the future path of the present does not need to be fixed. Such a model was endorsed by McCall, St. (1994). *A Model of the Universe*. Oxford: Clarendon Press. Cf. also Miller, K. (2005). Time Travel and the Open Future. *Disputatio I, 19*, 223–232.

[28] Minkowski, H. (1982). Raum und Zeit. In H. A. Lorentz/A. Einstein/H. Minkowski, (Eds.), *Das Relativitätsprinzip. Eine Sammlung von Abhandlungen (8th ed.)*. Stuttgart: Teubner, 54–66, p. 54: "Von Stund an sollen Raum für sich und Zeit für sich völlig zu Schatten herabsinken und nur noch eine Art Union der beiden soll Selbständigkeit bewahren."

[29] Einstein, A. (1973). *Über die spezielle und die allgemeine Relativitätstheorie (21st ed.)*. Berlin: Akademie Verlag, p. 96: "Die Physik wird aus einem *Geschehen* im dreidimensionalen Raum gewissermaßen ein *Sein* in der vierdimensionalen 'Welt'."

[30] Weyl, H. (1924/1977). *Was ist Materie?* Berlin: Springer, p. 87: "Die objektive Welt *ist* schlechthin, sie *geschieht* nicht. Nur vor dem Blick des in der Weltlinie seines Leibes emporkriechenden Bewußtseins 'lebt' ein Ausschnitt dieser Welt 'auf' und zieht an ihm vorüber als räumliches, in zeitlicher Wandlung begriffenes Bild."

[31] The parameter t is converted into an imaginary variable with spatial units by $\sqrt{-1}ct$.

[32] Reichenbach, H. (1955). Die philosophische Bedeutung der Relativitätstheorie. In P.A. Schilpp (Ed.), *Albert Einstein als Philosoph und Naturforscher* (pp. 188-207). Stuttgart: Kohlhammer, p. 204: "Die Zeit ist die Ordnung von Kausalketten. Das ist das außerordentliche Ergebnis der Einsteinschen Entdeckungen: Time is the order of causal chains. That is the extraordinary result of Einstein's discoveries."

[33] Riemann, B. (1990). Über die Hypothesen, welche der Geometrie zugrunde liegen. In R. Narasimhan (Ed.), *Gesammelte mathematische Werke, wissenschaftlicher Nachlaß und Nachträge. Collected papers* (pp. 304-319). Berlin/Heidelberg: Springer, p. 318: "... daß die Metrik des Raumes unabhängig von den in ihm sich abspielenden physischen Vorgängen festgelegt sei und das Reale in diesen metrischen Raum wie in eine fertige Mietskaserne einziehe".

[34] For a thorough analysis of the indispensable difference between necessary, factual and possible entities and events cf. Evers, D. (2006). *Gott und mögliche Welten. Studien zur Logik theologischer Aussagen über das Mögliche* (Religion in Philosophy and Theology 20). Tübingen: Mohr-Siebeck.

[35] Cf. Tooley, M. (1997). *Time, Tense and Causation*. Oxford: Clarendon Press.

[36] Cf. Markosian, N. (2004). A Defense of Presentism. In: *Oxford Studies in Metaphysics*. (Vol. 1, pp. 47-82). Oxford: Oxford University Press.

[37] Davies, P.D. (1974). *The Physics of Time-Asymmetry*. London: Surrey University Press, p. 109.

[38] Prigogine, I. (1978). Time, Structure, and Fluctuations. *Science, 201*, 777–785, p. 785.

[39] Cf. *e.g.,* Peirce (1992).

[40] Denbigh (1981), p. 166.

[41] Cf. Luther, M. (1982), *Studienausgabe vol. 2* H.-U. Delius, (Ed.). Berlin: Evangelische Verlagsanstalt, p. 296: "qua alteri gratis (et) sponte seruit" (De libertate christiana = Weimar Edition 7, 64).

[42]   Cf. Kierkegaard's notion of repetition.

[43]   Cf. Dalferth, I.U. (1997). Gott und Zeit. In *Gedeutete Gegenwart. Zur Wahrnehmung Gottes in den Erfahrungen der Zeit* (pp. 232–267). Tübingen: Mohr-Siebeck and Dalferth, I.U. (2006). *Becoming Present. An Inquiry into the Christian Sense of the Presence of God.* Leuven/Paris/Dudley, MA: Peeters.

[44]   Tillich, P. (1959). *Theology of Culture* R.C. Kimball (Ed.). New York: Oxford University Press, p. 213.

[45]   Cf. Foster, D. (2000), Tillich's Notion of Essentialization – A Preliminary Spreadsheet of Observations. In D. Lax / G. Hummel (Ed.), *Mystisches Erbe in Tillichs philosophischer Theologie / Mystical Heritage in Tillich's Philosophical Theology* (Tillich Studien 3), Münster: LIT, p. 365–386.

CHAPTER 10

Time and Eternity: The Ontological Impact of Kierkegaard's Concept of Time as Contribution to the Question of the Reality of Time and Human Freedom

Elisabeth Gräb-Schmidt[*]

Faculty of Protestant Theology, University of Tübingen, Germany

Abstract: Is time a matter of consciousness only or can it draw upon reality? Not only since Augustine's question "But then what exactly is time?" the status of time has been challenged, but also much earlier during the days of the Ancient Greek philosophers. They already realized that the question of time must be confronted with the thought of eternity. For them, it was without question that eternity is the transcendental background of time. The insights of modern physics force us to reconsider 1. the meaning of time and its status of reality; 2. the relationship between absoluteness and relativity of time; 3. the idea of an eternal universe and the eternity of time.

This paper raises the question of a possible combination of eternity and time. Only if both aspects can coexist, we will be able to keep the idea of human freedom. In spite of a common prejudice, an eternal universe and its deterministic natural laws do not interfere with freedom as long as time plays a substantial role in our general rules of operation. Moreover, determinism plays an even necessary role, because it is only within determinism that a rational concept of freedom in terms of following rules can be maintained. It is time, which saves the freedom of the human will in spite of and even because of the determinism of natural laws.

Therefore, eternity and determinism do not contradict the freedom of will. However, it deserves an ontological concept of time, which is founded on a concept of eternity, not being "timeless" but also bearing time.

Keywords: Time, eternity, Søren Kierkegaard, Christoph von der Malsburg, the ontological status of time, the eternal universe, contingency, determinism, human freedom.

1. INTRODUCTION

The special relation and interference of time and eternity could be traced back to

*Address correspondence to Elisabeth Gräb-Schmidt: Institute of Ethics, Faculty of Protestant Theology, University of Tübingen, Germany; Tel: +49 7071 2978023; Fax: +49 7071 295415; E-mail: elisabeth.graeb-schmidt@uni-tuebingen.de

the relation of evolution and emergence. They coincide in a special pattern of paradoxality, which functions as the transcendental background of reality itself.

Whoever is interested in an interdisciplinary workshop has at least one goal: to keep in touch with the firm conviction of the oneness of reality. One of the outstanding subject matters to discuss the possibilities of a unifying world view is the subject of time. Time is looked at in different ways – dependent upon whether it is dealt with in physics, philosophy, theology, psychology or in everyday life. Therefore we will have to scrutinize whether there are remaining conflicts between the different disciplinary concepts of time or whether there are only different aspects which could be linked together in order to create a coherent theory of time.

There are – roughly speaking – three positions concerning the status of the reality of time:

Time is an idealized measurement called the "objective time". This belongs either to a realm of absoluteness or is considered solely as a physical parameter without implying any ontological issues. Such objective time has to be distinguished from a subjective time, which has value on the psychological level only.

Time does not have ontological status. The experience of time is contingent, due to culture and is to be considered as a practical and technical habit only [1]. Therefore, we will have to leave behind any theological notion of a unified model of time and eternity and thus any ontological conception of time. Time has to be considered as contingency in radical fashion [2].

Time has ontological status. Within time, we become aware of an intensive portrayal of human reality. The profile of this portrayal is sharpened when there is a combination of natural sciences and natural philosophy to categorize our knowledge, our observations and reflections on time. Let us take into account the results of modern physics. Prigogine, for example, states with regard to the theory of irreversibility: "Whatever the future of these ideas, it seems to me that the dialogue between physics and natural philosophy can begin on a new basis. I don't think that I exaggerate by stating that the problem of time marks specifically the divorce between physics on one side, psychology and epistemology on the other. (…) We see that physics is starting to overcome these barriers" [3]. Beyond this, there are indications in modern physics that support the ontological universality of time [4].

For a long time, natural sciences and humanities seemed to occupy different realms, which were not compatible. The concepts of causality and determination did not fit into a concept of freedom as the essence of the human being. And, with regard to the subject of time, the concept of linear time in classical physics did not fit into the concept of time as subject to inner consciousness - or, religiously speaking, time as it is conceived in everyday life was opposed to eternity. The measurement of time in continuously following, self contained states is not the time which represents our everyday intuitive conception of time. In line with these distinctions, the neurobiological dispute about the possibility or impossibility of a free will points to the heart of the boundary between the different disciplines of natural sciences and humanities. Moreover, it appears to gain the authority of interpretation in the field of humanities as well. In addition, the neurological insight into the human brain seems to annihilate the free will, as famous German neuroscientists (*e.g.,* Wolf Singer [5] and Gerhard Roth [6]) are convinced.

Nevertheless, the good news is that those diametrically opposed theories are no longer state of the art. In both fields we can observe a shift of paradigm, which makes possible a fruitful dialogue. Since the beginning of the last century, modern physics has changed from classical mechanics with its firm conviction of causality and determination to the theories of quantum mechanics and relativity. From those models of reality a different view on causality and determination seems to be possible and therefore we will gain an intrinsic reason to re-establish the interdisciplinary cooperation, because from this time onwards we can discern similar objectives. Moreover, through this recently accomplished meeting-point, new perspectives of the discussion of free will in neurosciences seem to be possible. These will rearrange the parameters of causality and determination so that freedom itself attains a new outlook [7].

In the face of these new results, a higher level of interdisciplinarity is now possible. Much to my surprise, I stumbled across a striking similarity of a new concept of physics and neurosciences as outlined by Christoph von der Malsburg *et al.* as well as the concept of time in the works of the Danish philosopher and theologian Søren Kierkegaard.

First of all, I shall ask a basic question concerning every theory of time (2.). Secondly, I shall cite some remarkable points of view within modern physics (3.). Thirdly, I shall shortly outline the concept of time in Kierkegaard's thinking (4.). Fourthly, I shall confront (possible) results of the previously established new level of discourse where both disciplines may be capable of learning from each other. Furthermore, I will try to engage with new hints for further speculations (5.).

2. THE BASIC QUESTION

From the very beginning in Ancient Greek philosophy time was a preeminent philosophical topic. Thereby it was a principle to look at the relation between a somewhat objective time and time as a matter of appearance in consciousness. So, what we now sometimes consider to be new when we speak of the "temporalisation" ("Verzeitlichung") of time in Twentieth Century philosophy, *e.g.,* in Heidegger's thinking, refers to problems discussed on a primary level since the very beginning of philosophy in one way or another. The reason for us to forget this constellation might be the still prevailing paradigm of the solid world view in classical mechanics as a measure for what can be called "real" in a naturalistic sense. From this point of view we strictly keep apart a sphere of objective time from the sphere of subjective time, whereas "real" in a naturalistic sense belongs to the laws of physics only, and not to the mere subjective sphere of consciousness. Therefore, the deep question of how the different modes of time can be brought together could be eclipsed by means of the paradigm of that kind of time rendered by the natural laws of classical mechanics. This has given rise to an understanding of time as "number" only.

In Ancient Greek philosophy, this understanding of time as number was represented in the concept of "Chronos". But in addition to this, there was always the understanding of time as a dimension within experience. In order to bring both conceptions of time together, the objective understanding as number and the subjective understanding of experience were considered as a task for philosophical reflection correspondingly. However, whereas the "reality" of time was beyond doubt in ancient times, it is now called into question when time in physics is considered as a quantity and not as a quality, and as a quality solely for subjective consciousness [8]. Based on new possibilities of conscious time in physics, I am in search of a coherent model of time in order to reconcile our ideas of physical time and psychological time.

Hence, the further question concerns the reality of time in an ontological sense. Is it indeed just a contradiction between time and eternity or is it possible to understand both as complementary? How do we have to deal with the question of the reality of time? Is our everyday understanding of time, our intuitive outlook on time helpful in order to understand the reality of time or does it only distract our view?

Our intuitive outlook on time does justice to time by exhibiting a special emphasis on the present moment. Whereas the past, on the one hand, is not real any longer, but is present in our memory only and therefore not real as such, the future, on the other hand, is not yet real. Moreover, the past is considered to be an accumulation of erstwhile moments. Only the present seems to be real.

If we take this view of time for granted, it is then not astonishing that time was always looked at as something subjective and that its ontological sense was not recognized. This assumption, however, bears persistent difficulties and makes use of an objective reality of time, from which the subjective apperception cannot abstract. Without going into details of the physical understanding of time, however, when expanding on the question of the combination of past, present and future, it seems to be necessary to assert an ultimate reality of time. And above all, we have to question how and why the present moment always is emphasized in the intuitive outlook on time and how, despite this special emphasis on the present moment, we may still arrive at an equally valued meaning of past, present and future. It is this question, which is a cornerstone for the claim of the ultimate reality of time, because it is only when present past and future are equally valid, we can pledge for time as an objective reality.

Now we are just getting some hints from modern physics for the plausibility of such an objective reality of time. Christoph von der Malsburg, a physicist and neuroscientist, is oriented towards an understanding of time, which is capable of including our understanding of time in everyday life. In addition to this, his concept of time is capable of including the results of modern neurobiology, which will be discussed later on. In the second part, I shall briefly sketch his thoughts out of two reasons: First of all, I believe that his theory is worthwhile being noticed as such. Secondly, his theory has a striking similarity to that model of time, which has been developed 170 years earlier, namely by the Danish philosopher Søren Kierkegaard. Therefore, this consonance of the two models may deliver an informative basis in order to put forward a coherent model of physical and psychological time.

3. THE SHIFT OF PARADIGM IN MODERN PHYSICS AND IN THE NEW OUTLOOK ON TIME [9]

In classical mechanics we deal with an idealised conception of time, which is considered to be linear and reversible. Future and past are equally predictable and therefore the time bar is not an arrow, not directed and reversible. The laws in classical mechanics are invariant with regard to time.

These idealistic constructions were questioned by physics itself as a result of the discovery of the second main theorem of thermodynamics. It has thereby been shown that the time bar is not reversible but has to be conceived as an arrow. Physical time as compatible with the law of thermodynamics no longer is reversible but remains irreversible. Such conception initiated the question of the reality of time and whether we have to deal with a beginning of the universe or not, once again. In fact, we have to consider that quantum theory in a fundamental sense also is time-reversal invariant as it is in classical mechanics. So it seems to be a single discovery in thermodynamics, yet of such kind that it still causes severe speculation about ontological questions and that it brings philosophy and physics closer to one another. As a consequence, time becomes a crucial factor for posing the question of the status of reality of physical laws and thus for opening up the possibility of combining the laws of physics with the reflections of metaphysics. Although time in quantum mechanics and in the theory of relativity was not dependent upon the conception of irreversibility belonging to thermodynamics, the idea of an absolute time and an absolute space was flawed nevertheless. The peculiar status of the observer and his relation to space and the speed of light shattered the conviction of an ultimate reality of objective time and space. But still, to consider the status of time did not seem to be of particular urgency for studying the laws of these theories. Nevertheless, time had an impact with regard to the consequences. This impact could be discerned in the ongoing conception of the chaos theory and its consequence of indeterminism. It is obvious that the chaos theory takes the impact of the irreversibility of time shown in thermodynamics seriously, while it is still keeps in touch with classical physics.

For this new point of view opened up by chaos theory, time asserts the asymmetry of thermodynamics by pointing out that physical laws seem to be directed towards the future. Therefore, the spectacular phenomenon consists in the fact that quantum effects always end up in the future [10].

These results of quantum mechanics are thus not fully taken into account within the traditionally conceived physical concept of time, as Christoph von der Malsburg notices. Two aspects of these results are missing:

1.   the equal reality of the past, the present and the future.

2.   the priority of the future.

These aspects have to be reconciled with a new theory or a more accurate interpretation of time. Von der Malsburg realizes that, in fact, his reflections are not shared by most physicists. Furthermore, they heavily oppose our intuitive outlook on time and life. However, in light of our concepts of mind processes, it leads to remarkable possibilities when we look at time in a holistic way. The key point of this new view on time is that there are arguments within physics in favour of an eternal universe. The full dimension of time has a simultaneous reality. If the word "simultaneous" is not taken to refer to the same "t" but is taken to mean that physical processes at any moment of the universe can be influenced directly by physical processes at any other moment, that is present, past or future, then this leads to the conception of an eternal universe [11]. Yet, the very interesting point of von der Malsburg's claim consists in the fact that – in contrast to first sight – this eternal universe does not lead to a deterministic conception of reality where everything is already fixed. Rather, in light of our new concepts of mind processes, it gives room for the idea of free will. Even though a universe with eternal reality appears to render a fixed frame, it is indeed due to this concept that we can postulate human freedom in an eminent sense.

First of all, I will very briefly sketch the arguments for an eternal universe, which von der Malsburg draws from the constructions of quantum mechanics and the theory of relativity. Secondly, I will point out the consequences with regard to the meaning and the reality of time. Thirdly, I will compare these results with the conception of time as we find it in the writings of Kierkegaard. For the latter it provides the cornerstone of his view on the human mind, the soul and faith, having both theological and anthropological bearings.

Von der Malsburg follows two stages:

1. He makes the case for an eternal universe by means of arguments taken from physics.

2. He draws the consequences from an eternal view of the universe for the meaning and the reality of time.

Ad 1 Physical Arguments for an Eternal Universe

The intuitive outlook on time conceives of the reality of time in terms of the present moment. This understanding of time is reflected in field physics, since within this discipline full reality always is presented in terms of a time-slice, which contains all signals from past events and thus contains all information necessary in order to produce the next moment. History of the universe is rendered in an incremental way, that is, as a wave of real moments. Even though, however, such interpretation of time corresponds with our intuitive notion of time to a great deal, it nevertheless leads to problems when it comes to a confrontation with the results of relativity on the one hand and quantum physics on the other hand. According to relativity, one past moment "t" is different for different observers. If we then claim equal status for different observers, we will nevertheless end up with the concept of a "simultaneous reality" [12]. Furthermore and above all, the model of the time-slice is problematic within the realm of the quantum phenomena, because of the necessity of non-local-communication. It seems to be the very model of a time-slice, which has to be reconstructed under the new conditions of the theory of relativity and quantum mechanics in order to keep track with reality. Von der Malsburg argues that because the emitted wave of a photon reaches many potential absorbers and nevertheless is absorbed in one place only, the interference between the potential absorbers is necessary. Thus, von der Malsburg's concern is a more accurate interpretation, which does not imply the figment of a projection operator leading to further problems only, such as, to mention the worst, foggy arguments in favour of a distinction of "real" and "in the mind" [13]. Correspondingly, his main aim concerns the reconciliation with a system of interference between potential absorbers in which we are not operating with invoked and retarded signals only, but in which we are dealing with advanced signals [14]. Everything is organized by some kind of "handshake" involvement [15]. In addition to the ordinary and retarded wave, an advanced wave is proposed, namely one that follows the retarded wave backwards in time as a kind of "confirmation wave" of the one emitter [16]. Von der Malsburg assures that no conflict with relativity is rendered, since any interference takes place along the light cone and no concepts of simultaneity will ever be invoked [17].

Ad 2 Consequences for the Meaning and the Reality of Time Flowing from an Eternal View of the Universe

For von der Malsburg, this concept hints at the full reality of all moments of time, *i.e.,* of past, present and future [18]. However, not only the full reality of all segments of time is evoked, but also a peculiar priority of the future is invoked due to the quantum phenomena. And this is remarkable. Although it seems to be the case that the nature of transactions in quantum mechanics is symmetrical with time, as the natural laws in classical physics are, quanta always end up in the future, never in the past [19]. He concludes that we have to deal with some kind of forward causation [20]. Hence, the exciting conclusions, which von der Malsburg draws, are:

1. The creation of the universe cannot be described in terms of operating with a wave progressing from a moment of creation to the present and

finally to later moments. At the birth of the universe, its end was present already [21]. He concludes:

2. The universe is an eternal "entity" being completely static "rigid as a crystal, metaphorically speaking" [22].

Considering the universe as a rigid, complete entity can be related to the notions of determinism and indeterminism, conceived as a choice between different possibilities. It operates with time as being integrated into a fixed structure of determinism. Within field equation, however, it can be put in terms of probabilistic predictions only.

But as a result of these reflections, "an eerie feeling of living on theatrical stage" [23], of replaying for what has happened already remains. An important issue for von der Malsburg is the structural stability of the world with the aim of reconstructing a theory that takes into account consciousness. Therefore it has to be symbolic, an exact temporal replica [24]. We have thus got to conclude that we are not capable of perceiving time directly, but in a symbolic and indirect way only. If this symbolic theory of time is compatible with physics and our intuitive outlook on time, there will be no clash with the idea of an eternal universe whatsoever. Determinism in this picture is not in control of the progression of time, as it is usually conceived, but refers to a completely rigid array of retarded and advanced signals. "It is, metaphorically speaking, as if the universe had already run its whole course and we were dealing with a recording" [25]. But this should not bother us. Von der Malsburg points out that to some extent, determinism is indeed necessary for dealing with predictabilities as for our conduct in everyday life as well as in construing our theories [26]. In the same way, we are not able to falsify the existence of physical laws in our actions.

But what does happen to freedom then? If determinism is true, is it still possible to hold on to the conviction of a free will? Von der Malsburg confirms that within the conception of a forward causality, there is still room for interaction. My actions do influence the future, but what I do change is not the real future but the potential future of my imaginations and predictions [27]. Therefore, this determinism has nothing to do with neglecting a free will. Because a "deterministic and an eternal universe would be an unsupportable prison only, if we and our mind mechanism weren't an integral part of it. But as the universe obeys forward causality, we are very concretely participating in forming the future" [28]. Therefore, we have to take the "action principle" [29] serious, which deals in a free way under the signal of forward causality and thus within the sphere of our action we are able to shape the future not instead but because it is determined. This makes sense against the background of the requirement that the universe is eternal and "already there" and that we have to deal with the recording – and this has to be done as well as it displays the room of our freedom.

To my surprise, this concept of the relation of eternity and time together with the peculiar theory of freedom within determinism, putting emphasis on the action principle in an eminent way, can be directly conferred to the concept of time as Søren Kierkegaard construes it.

It is the symbolism of time which delivers the model of a coherent theory and which, in addition, verifies the claim of an ultimate reality of time. This symbolism corresponds to Kierkegaard's theory with the parameters of "Zeitlichkeit" (temporality), "Wiederholung" (repetition) and "Vorsehung" (providence). All three terms describe our relationship with time and eternity as a symbolic form.

4. THE CONCEPT OF TIME IN SØREN KIERKEGAARD'S THINKING

Kierkegaard did not write a self-contained treatise on time. His theory must be extracted from his different writings (*Sickness unto Death*, *The Concept of Anxiety* and others [30]).This does not mean that time plays an unimportant role in his thinking. On the contrary: Because it does play a significant role, time is mentioned throughout the whole of his work. It is the pivotal point of the conception of human existence. Long before the explanations of Heidegger's *Being and Time* [31], it is Kierkegaard who invents the term "Zeitlichkeit" [32] to describe the reality of the human being and, above all, of reality as such. In fact, he has anticipated the phenomenological theories of time of Husserl [33] and Heidegger.

To elaborate on this would not be worthwhile within the constraints of our context. However, Kierkegaard's work is worth mentioning, because he does not only describe the inner consciousness of time and therefore the aspect of the subjectivity of time in detail. He also raises ontological questions and establishes the objectivity of time anew at the same time. Due to this, the reflections of Kierkegaard are compatible with those of von der Malsburg. In virtually the same way, Kierkegaard envisions connections to eternity and also to determinism and freedom, which probably provide a basis for the objectivity of time. In order to illuminate those correlations, I shall first refer to Kierkegaard's evaluation of the linear concept of time and the notion of the present moment within everyday life. Thereafter, I shall refer these to their possible relation to eternity. This clarifies the role of the determinism in its – theologically and anthropologically important – possible relation to freedom.

A central aspect of his theory of time is the conception of "Zeitlichkeit" (temporality). "Zeitlichkeit" refers to exactly that relation between life-time and universal-time and therefore is capable of relating the psychological with the physical time as well. Thus, this relation may bridge the gap between the subjective notion of time and what is called objective time.

4.1. The Present Moment

In order to understand Kierkegaard's concept of "Zeitlichkeit", we have primarily got to turn to his notion of the present moment and his relation to the physical understanding of time on the one hand and to eternity on the other hand. The present moment actually is the crucial category in order to qualify the understanding of time. It is the foundation of the past, the present and the future [34]. It is the eternity within the present moment, which can be looked at as the very foundation of time. And this means that the quality of time does not rest in time itself, in its linearity, but in the reference to eternity [35]. The "now" of the present moment is not time itself. On the contrary, it is beyond time, but in exactly this way, it constitutes time as succession [36].

It is this conception of time, which Kierkegaard brings to light against the reality claim of time being just linearity. The concept of the linearity of time fails to exhibit the full reality, because it does not take into account the role of the present moment, according to Kierkegaard [37]. As a foundation of time, Kierkegaard does not understand the present moment as an atom as it is understood in the philosophy of Parmenides, not as the sudden, but rather in terms of the "νυν" in Aristotle's theory of time. Kierkegaard does refer – as Aristotle does – to the "Janus-facedness", which in fact cannot explain the transition from one mode of time to the other, but which always envisions them by means of dividing the past from the future. The "now", the present moment, does thus, in fact both: it does constitute the linearity in the sense of the accumulation of time-slices as well as it does qualify time. Correspondingly, the present moment plays a double role in Kierkegaard's thinking: first it guarantees the continuity in the sense of the Aristotelian **"νυν"** in that it shows the transition from past to future through its Janus-facedness [38]; secondly it emphasizes the present moment as such in its relation and depiction of eternity. It emphasizes the present in its qualitative counterpart to past and future. It is this qualifying function of the present moment that leads to the conception of "Zeitlichkeit".

Let us summarize: The present moment displays a twofold function. On the one hand, it is subsumed under the linearity of time by dividing the time-slices. On the other hand, it is singled out in its punctual timelessness. As such, a timeless point within the time-flow constitutes time in the direction of the different modi of time, the past, the present and the future and is thereby also qualifying time as "Zeitlichkeit", which emphasizes the present time. And in both cases, the moment is actually not part of time. Under the condition of succession, the present moment as a singular point of the time scale is not just time, but timelessness. Furthermore, in its function as **"νυν"**, *i.e.,* as dividing past, present and future, it emphasizes the present time, and the moment is thus distant from time. For Kierkegaard, this timelessness of the moment stems from its coming out of and its reference to eternity.

The double function of the present moment as constituting the continuity of time as well as accentuating the present time raises the question of the ontological impact of time no less than the question of the role of contingency and determination of the concept of eternity. The "now" in its function to render continuity calls for being saved in eternity. Without regard to eternity, the "now" in its particularity would be lost in a never-ending repetition of the same [39]. Being qualified by eternity, the moment in its presence does keep the reference to the past in light of the future. The ongoing moments, then, display not repetition as mere retrieval but as a "fully-fledged" activity, which always implies creative newness. The "now" as the external foundation of time, is not seizable in its quality within a one-dimensional, linear understanding of time and is not time in itself. However, it bestows endurance and qualitative distinctness upon time. The present moment does render this quality in virtue of its origin in eternity [40].

4.2. The Interdependent Role of Contingency, Determinism and Freedom in the Light of Time and Eternity

Contingency does indicate the not calculable, the unpredictable character of the future. In the present time, we become aware of it in the event of what happens in the present moment. This character of the present moment as an event can be interpreted in two different ways. Either it comes out of the blue without involving any rules at all, or it comes out of a higher level of order and law, out of a – so to speak – "objective reality". Such alteration cannot be decided in virtue of calculations or reflections, but in virtue of experience only – otherwise it would not be an event. Contingency exhibits a constitutive function in the "Werden der Zeitlichkeit". "Zeitlichkeit" can be explained as the breaking out of the linearity of time, being exposed to a peculiar impression of a contingent event that shapes the present moment. As this disconnection of linearity, it is definitely opposed to determinism in that it does not follow our rational rules of causality. But this does not mean that there is no determinism at all. It is only opposed to determinism in the sense of a calculable necessity and causality. On a higher level there may still be a place for some higher kind of determinism. But such higher mode of determinism is not related to necessity and causality but to contingency. Its relation to contingency still allows the thought of freedom. It could be conceived as being localized within the realm of eternity. At the level of life-time it will still allow freedom. And it is this kind of freedom in time by means of contingency that is founded on eternity and that Kierkegaard calls "Zeitlichkeit". "Zeitlichkeit", therefore, is only made possible by means of a peculiar interference of eternity with time and thus in freedom. Simply because time is bound to eternity, contingency does not mark fate but freedom.

At this stage, the interesting point is that it also goes the other way around: Because eternity is bound to time, determinism does not mean necessity but freedom. The character of contingency as freedom thus depends precisely on the interdependence of time and eternity, which makes determinism compatible with freedom.

"Zeitlichkeit", that is contingency, is a result of the point of contact between time and eternity. In this fulfilled moment we experience freedom as an open range of a time being conditioned by eternity. It is exactly this reference to eternity within the appropriation as present, which Kierkegaard labels "Zeitlichkeit". Kierkegaard makes this more precise by means of the following thought: "Zeitlichkeit" in the eminent sense conceives time as the "fullness" of time [41]. And this fullness does not mean the flow of time as a steady linearity, but as "full objectivity" of presence in the moment. Standing in this perspective we can realize that for Kierkegaard the process of change ("Werden") has a peculiar character. In every case, the corresponding change ("Werden") does not refer to an evolutionary change ("Werden") a collation of sequences of states in the sense of "time-slices". As "change" ("Werden") it indeed tends towards the future, but this future in no way is a linear prolongation of the past and presence. Rather, reality as "change in time" ("Werden in der Zeit") is – so to speak – an eschatological qualification of the presence just through its reference to eternity and being under revision by the past [42].

With this background, we can realize that the thought of eternity and determination is not in danger of interfering with the idea of freedom, because from this perspective freedom will always appear as an opportunity amongst a variety of possibilities. And these possibilities are always possibilities of an appropriation of what already is there (in eternity). And there is no other way but through this variety of appropriation that something new arises, that is, creativity. Therefore, freedom arises through a variety of appropriations, according to Kierkegaard. Freedom is concentrated in the various manners of appropriation of reality. Hence eternity and freedom do not contradict one another. Furthermore, time is not levelled by eternity. On the contrary, time not only is important with regard to our ordinary life, but also with regard to subjective consciousness with its "für uns". It also has value for the figure and profile of eternity itself. That is, eternity is not insensitive to the realisation of possibilities [43]. And it is here that we can discern the main difference to a traditional deterministic world view. The reference to eternity bears substantial consequences for the progress of our realisations, but it is not a fixed recording. Rather, the meaning of time as time is qualified by means of the reference to eternity as freedom, on the one hand, and a dependent order, on the other hand. Therefore, it is the presupposition of an eternal determination that allows both rational legislation/causality and freedom within the realm of time.

The moment, the "now", qualifies time in a double way: On the one hand, the moment leads to the qualification of time through the perspective of eternity. Here the moment is conceived as the fullness of time, as "καιρός". On the other hand, it leads to an unauthorized reality when the moment is nothing but an ever vanishing and ever repeating point on the time scale. For Kierkegaard, this perception of time leads to non-existence (Nichtsein). Hence, reality is constituted as an open-ended change ("Werden"), metaphorically speaking as a "leap", a "plunge", condensed in the "now", which in temporal terms means "nothing", *i.e.,* which is not time itself [44]. And this exactly exhibits the attribute of the "now" in Aristotle. However, it does have the power of structuring time in both ways, be it as eternity as including and qualifying the succession of time, or be it as linear succession under exclusion of eternity and thus solely quantifying but not qualifying time. Correspondingly, time gains its quality through the structure of a relatedness in virtue of which a variety of possibilities is realized through simultaneity of time and eternity in the "now" (fulfilling moment).

The key point of a Christian understanding of time compared with a Greek Platonic understanding is that eternity does not only qualify time - or that eternity does not exclusively rest in eternity but also in time -, but that time is qualifying eternity as well.

4.3. Comparison of the Conceptions of Time in Søren Kierkegaard and Christoph von der Malsburg

The relation of time and eternity as depicted in the conception of von der Malsburg can be easily related to the respective conception I have outlined with regard to Kierkegaard. In particular, this becomes apparent when we focus on the Christian dogmatic theme of providence. In virtue of the extension of time to eternity, eternity does no longer have to be understood in terms of determinism but in terms of possibility. Thus, eternity makes room for open creativity. Such an understanding of a definite eternity including possibility and thus freedom was developed in the Christian doctrine of providence. The Christian conception of providence does not mean determination in a closed sense but guidance. The respective "change" ("Werden") being guided by contingency has to be understood as a new change ("Werden") in time through the category of eternity as providence. This gives account of the fact that we are not dealing here with a predetermination of our actions but with a relational preordination, a relational connecting track of what happens (see also chapter 4). Since it is always the case that reality stems from time and eternity, it is the reference to eternity that guarantees objective reality and it is the reference to time that allows a dynamic shape of universal reality. It is exactly this dynamic reality, an ever "being in change" ("Sein im Werden") of reality that Kierkegaard emphasizes in his theory of time not only with regard to temporality but also with regard to eternity.

The combination of time and eternity reminds of Kierkegaard's ontological interest, which he shares with von der Malsburg. And as the replay of the seemingly fixed recording of the universe in the concept of von der Malsburg still allows for freedom, the fundament of time in eternity and the shape of eternity through time in Kierkegaard does allow for freedom as well. Therefore, this relation of time and eternity does always occur by means of the category of contingency. Such contingency is precisely that moment, which interrupts the subjective sphere of time. With contingency, time cannot be just reduced to a phenomenon of consciousness. Kierkegaard does not intend to relativize temporality in favour of eternity. Rather, he intends to qualify – as is the case in von der Malsburg – the outer sphere of physical time. Contingency, however, is precisely what enters from the outside into the inner sphere of consciousness. Therefore, contingency refers to the ontological sphere in contrast to a subjective meaning of a somehow constructivist view of reality only. Viewed from his theory of time, the loss of contingency indicates the loss of eternity in time and thus of the ontological background. The contingency as a category of eternity can be glimpsed in time within the present moment.

The structure of human freedom being under the rule of an eternal providence is coined "repetition" ("Wiederholung") by Kierkegaard [45]. We can compare it with the peculiar mode of action in von der Malsburg's view of the universe as a fixed recording – but with the inclusion of freedom. The difference between the present moment as something continuously reappearing in infinite repetition and infinite succession on the one hand, and a qualitative temporality qualifying the "now" of the moment in terms of eternity on the other hand, becomes apparent due to the reference of contingency to eternity. Based on this reference, the present moment is not situated in a one-dimensional time scale. It is rather due to its foundation in eternity that repetition acquires its peculiar character. Such repetition does always indicate a kind of growing in change ("Werden"), a creativity. It embraces a reference to what has already been, yet this reference contains a moment of appropriation, which regains the past and thus changes it. Such growing through appropriation is made possible due to the structure of the present moment as the simultaneity of continuity and contingency. Simultaneity precisely refers to the extension of eternity into temporality. Repetition, therefore, denotes the conception of qualitative temporality due to the coincidence of time with eternity. Temporality is qualified by virtue of eternity, but it does nevertheless involve our actions and thus exhibits freedom within a predetermined order. Such eternal qualification coexists with a temporal continuity. The former refers to the present moment as qualitative time condensed in the "now", which is not time itself but belongs to the eternal order, whereas the latter refers to natural necessity.

5. THE UNDERSTANDING OF REALITY BASED ON AN ONTOLOGICAL MODEL OF TIME

The status of the theory of time in Kierkegaard's thinking can be considered from the viewpoint of the consequences it bears for our understanding of determinism and freedom. It is very important that the consequences of an ontological theory of time and eternity do not end up in determinism. This is the aim for both, von der Malsburg and Kierkegaard.

The relation of eternity and time being concentrated in the present moment is relevant for the understanding of reality. The present moment embodies a paradigmatic function in the presence of contingency. It is a single moment in time, which is at the same time beyond time and it is precisely this position that is constitutive of time as temporality. Therefore, the present moment is capable of preserving the former in the face of future possibilities and of making it available for appropriation. It is such appropriation that renders repetition – "Wiederholung". It is as "Wiederholung" only, as repetition in a qualitative sense that something new is called into being, according to Kierkegaard [46]. And while such newness has ontological severity only, it acquires reality through repetition only. For such kind of novelty, eternity is not made superfluous, as it is in the relation of time and eternity only that reality emerges.

Because of the constitutive function for time and eternity, contingency and not necessity provides the background and the origin of being. It is precisely this, which signifies the structure of "being as change" ("Sein als Werden") – or of a replay of a fixed recording. We have to take into account a peculiar interplay of time and eternity. As the singularity of the present moment, the general term of eternity attains an ever new face, and this is the reason for the fact that contingency exhibits a decisive, indeed a constitutive function for being, for reality itself.

I have already mentioned the word "providence" [47]. Providence gives account of a qualified understanding of eternity – qualified by its relation to time excluding a closed determinism. Or, providence moulds a peculiar shape of reality. Such shape of reality through providence gives a reality being guided by rules but still giving room for openness of such kind that it still follows some order. We are not dealing here with chaos but with a well-defined structure. This structure is retained in virtue of the role of the present moment being situated in both realms, the "open" time and the predetermined eternity. But in its coexistence the present moment is contingent. As such the present moment does always indicate the realisation of open possibilities, which are possibilities within a cosmic order, but which are contingent in their punctual realisations.

In Kierkegaard's thinking, the coexistence of contingency and determinism becomes possible in virtue of a peculiar understanding of a deterministic order as providence. Providence means guidance and is safeguarded due to temporality having its roots in eternity. As I have already mentioned, this takes into consideration the fact that eternity does not only qualify time, but that time as qualitative temporality does qualify eternity as well, this being due to an understanding of eternity as providence. Providence holds eternity off a closed determinism. Rather, it renders freedom, that is, our freedom consisting in this kind of guided openness of the freedom of eternity and its peculiar open order. It is due to this only, that natural legislation and freedom can be seen together without contradiction. This openness indicating freedom carries the peculiar direction towards the future. It does not just occur as an ordinary state of affairs, but it is an occurrence of a peculiar kind. Correspondingly, reality can be described as intentionality, being motivated in virtue of the interference of temporality and eternity in the process of the constitution of linear time as well as the peculiar emphasis on the present time.

Therefore, the theory of time as put forward in Kierkegaard's thinking, indicates that our natural order, our historical and natural boundaries, do not constrain our freedom but indeed establish it. Freedom exists as intentionality appropriating this order. And, to come back to the theory of von der Malsburg, this appropriation refers directly to his idea of the replay of a fixed recording [48]. In fact, it is in need of a legislative order - as, for example, being displayed in natural laws-, in order to then being appropriated deliberately. And therefore, natural law, "authentic" order, does not contradict freedom. Rather, it is freedom that does presuppose such order.

At this point, another aspect in von der Malsburg's conception does recur in Kierkegaard's thinking: It is the principle of action. This principle lies at the heart of the idea of intentionality. In Kierkegaard's thinking, it is the cornerstone of the temporal structure of human existence. Furthermore, it is the very cornerstone of reality as such.

ACKNOWLEDGEMENTS

Declared none.

CONFLICT OF INTEREST

The author(s) confirm that this chapter content has no conflict of interest.

REFERENCES

[1]   Rorty, R. (1989), *Kontingenz, Ironie und Solidarität*, Frankfurt a. M., p. 306.

[2]   Rorty, R., *Kontingenz, Ironie und Solidarität*, p. 50.

[3]   Prigogine, Ilya (1973), 'Time, Irreversibility and Structure', In: *Physicist concepts of nature*, ed. Jagdish Mehra, Dordrecht/Boston, p. 590f.

[4]   Prigogine, Ilya / Stengers, Isabelle (1981), *Dialog mit der Natur, Neue Wege naturwissenschaftlichen Denkens*, München, p. 267ff., p. 287f.; see also Wheeler, John Archibald (1979), 'Frontiers of Time', In: *Problems in the foundation of physics*, ed. Guiliano Toraldo di Francia, Amsterdam/New York, pp. 395-497, p. 407ff.; Lübbe, Hermann (1992), *Im Zug der Zeit, Verkürzter Aufenthalt in der Gegenwart*, Berlin/New York, p. 31.

[5]   Singer, Wolf (2002), *Der Beobachter im Gehirn*, Frankfurt a.M.

[6]   Roth, Gerhard (2000), *Das Gehirn und seine Wirklichkeit*, Frankfurt a.M.

[7]   von der Malsburg, Christoph (2003), 'Physics and our outlook on time', In: Albeverio, S. und Blanchard, P., *The direction of Time. The Role of Reversibility/Irreversibility in the Study of Nature*, Cambridge University Press, Cambridge, 1-14.

[8]   However, calling into question the reality of time might be due to a lack of clarity concerning the natural and material traces of the inner personal experiences of time. To deal with the material basis of those experiences leads to neurosciences and will have consequences to the problem of freedom and the free will. Whether those reflections could be reproduced in terms of physics will be a further question which cannot be answered in this paper.

[9]   For the following interpretation see [7].

[10]   See [7], pp. 2-4.

[11]   *ibid.*, p. 4.

[12]   *ibid.*, p. 5.

[13]   *ibid.*, p. 6.

[14]   *ibid.*, p. 3.

[15]   *ibid.*, p. 3.

[16]   *ibid.*, p. 3.

[17]   *ibid.*, p. 3.

[18]   *ibid.*, p. 2.

[19]   *ibid.*, p. 4.

[20]   *ibid.*, p. 5.

[21]   *ibid.*, p. 5.

[22]   *ibid.*, p. 5.

[23]   *ibid.*, p. 5.

[24]   *ibid.*, p. 7.

[25]   *ibid.*, p. 9.

[26]   *ibid.*, p. 9.

[27]   *ibid.*, p. 9.

[28]   *ibid.*, p. 12.

[29]   *ibid.*, p. 12.

[30]   In the folllowing analysis of the time concept of Søren Kierkegaard, I will mainly refer to his concept of Anxiety: Kierkegaard, Søren (1967), *Der Begriff Angst*, ed. Lieselotte

Richter, Reinbek; some of the other works, which are dealing with the concept of time are: Kierkegaard, Søren (1975), *Entweder-Oder I*, ed. H. Diem und W. Rest, München; Kierkegaard, Søren (1966), *Die Krankheit zum Tode*, ed. Lieselotte Richter, Reinbek; Kierkegaard, Sören (1964), *Philosophische Brocken*, ed. Lieselotte Richter, Reinbek.

[31] Heidegger, Martin (1979), *Sein und Zeit*, 15th edn, Tübingen.

[32] Kierkegaard, Søren (1967), *Der Begriff Angst*, ed. Lieselotte Richter, Reinbek, p. 82.

[33] Husserl, Edmund (1928), *Vorlesungen zur Phänomenologie des inneren Zeitbewusstseins*, Tübingen.

[34] Kierkegaard, Søren, *Der Begriff Angst*, p. 82.

[35] *ibid.,* p. 79.

[36] *ibid.,* p. 81.

[37] *ibid.,* p. 82.

[38] *ibid.,* p. 83, ref.1.

[39] *ibid.,* p. 83, ref.1.

[40] See also my interpretation of S. Kierkegaard's concept of time in: Gräb-Schmidt, Elisabeth, 'Leben aus der Vergangenheit', in: Cappelørn, N.J. *et al.* ed., 'Subjektivität und Wahrheit/Subjectivity and Truth', *Kierkegaard Studies: Monograph Series 11*, *Schleiermacher-Archiv 21*, 687-707.

[41] Kierkegaard, Søren, *Der Begriff Angst*, p. 76f.; p. 86.

[42] *ibid.,* p. 16.

[43] *ibid.,* p. 142f.

[44] *ibid.,* 16f; p. 76f.

[45] *ibid.,* p. 83, ref.1.

[46] See also: Kierkegaard, Søren (1966), *Die Wiederholung*, ed. Lieselotte Richter, Reinbek; cf. Beyrich, T. (2002), *Ist Glauben wiederholbar?*, Kierkegaard Studies, Berlin/New York.

[47] Kierkegaard's reflections on providence are displayed in: Kierkegaard, Søren (1966), *Die Krankheit zum Tode*, ed. Lieselotte Richter, Reinbek.

[48] See [7], p. 2.

LITERATURE

Baumgartner, Hans Michael (Ed.) (1996), *Das Rätsel der Zeit*, Philosophische Analysen, Freiburg

Beyrich, Tillmann (2002), *Ist Glauben wiederholbar?*, Kierkegaard Studies, Berlin/New York

Böhme, Gernot (1974), *Zeit und Zahl, Studien zur Zeittheorie bei Platon, Aristoteles, Leibniz und Kant*, Frankfurt a.M.

Gimmler, Antje *et al.* (Eds.) (1997), *Die Wiederentdeckung der Zeit*, Darmstadt

Gräb-Schmidt, Elisabeth, 'Leben aus der Vergangenheit', in: Cappelørn, N.J. *et al.* (Eds), *Subjektivität und Wahrheit/Subjectivity and Truth*, Kierkegaard Studies: Monograph Series 11, Schleiermacher-Archiv 21, 687-707

Heidegger, Martin (1979), *Sein und Zeit*, 15th edn, Tübingen

Heidegger, Martin (1989), *Der Begriff der Zeit*, ed. H. Tietjen, Tübingen

Husserl, Edmund (1928), *Vorlesungen zur Phänomenologie des inneren Zeitbewusstseins*, Tübingen

Kierkegaard, Søren (1964), *Philosophische Brocken*, ed. Lieselotte Richter, Reinbek

Kierkegaard, Søren (1966), *Die Krankheit zum Tode*, ed. Lieselotte Richter, Reinbek

Kierkegaard, Søren (1966), *Die Wiederholung*, ed. Lieselotte Richter, Reinbek

Kierkegaard, Søren (1967), *Der Begriff Angst*, ed. Lieselotte Richter, Reinbek

Kierkegaard, Søren (1975), *Entweder-Oder I*, ed. H. Diem und W. Rest, München

Lübbe, Hermann (1992), *Im Zug der Zeit*, Verkürzter Aufenthalt in der Gegenwart, Berlin/New York

Poser, Hans, *Zeit und Ewigkeit. Zeitkonzept als Orientierungswissen*, in: Baumgartner, Hans Michael (Ed.) (1996), *Das Rätsel der Zeit, Philosophische Analysen*, Freiburg, 17-50

Prigogine, Ilya / Stengers, Isabelle (1981), *Dialog mit der Natur, Neue Wege naturwissenschaftlichen Denkens*, München

Rorty, R. (1989), *Kontingenz, Ironie und Solidarität*, Frankfurt a. M.

Rorty, R. (1995), 'Philosophy and the future', in: *Rorty and Pragmatism*, ed. Hermann J. Saatkamp, Nashville, London

Roth, Gerhard (2000), *Das Gehirn und seine Wirklichkeit*, Frankfurt a.M.

Sandbothe, Mike (1998), *Die Verzeitlichung der Zeit, Grundtendenzen der modernen Zeitdebatte in Philosophie und Wissenschaft*, Darmstadt

Singer, Wolf (2002), *Der Beobachter im Gehirn*, Frankfurt a.M.

von der Malsburg, Christoph (1997), 'The coherence Definition of Consciousness', In: Ito, M./ Miyashita,Y./ Rolls, E.T. (Eds.), *Cognition, Computation and Consciousness*, Oxford University Press, 193-204

von der Malsburg, Christoph (2002), *How are neural Signals related to each other and to the world?*, J. Consiousness Studies 9,, 47-60

von der Malsburg, Christoph (2003), 'Physics and our outlook on time', In: Albeverio, S. und Blanchard, P., *The direction of Time. The Role of Reversibility/Irrreversibility in the Study of Nature*, Cambridge University Press, Cambridge, 1-14

Wheeler, John Archibald (1979), 'Frontiers of Time', In: *Problems in the foundation of physics*, ed. Guiliano Toraldo di Francia, Amsterdam/New York, 395-497

Zimmerli, Walther / Sandbothe, Mike (Ed.) (1993), *Klassiker der modernen Zeitphilosophie*, Darmstadt

CHAPTER 11

The Timelesness of Eternity from a Neuroscientific and Trinitarian Perspective

Wolfgang Achtner[*]

Institute of Protestant Theology, Justus Liebig University, Giessen, Germany

Abstract: This paper addresses three issues. In the first part the relation between consciousness and time is being discussed as it developed in the history of philosophy and theology. It covers Plato, Plotinus and St. Augustine. It continues in the second part to describe that time it is being perceived in the mystical consciousness as eternity which means in this context timelessness. Examples from world religions are offered. The question is asked if this eternity in mystical experience can be understood as relating to the eternity of God or as a mere self-experience. In order to settle this question mystical experiences as being interpreted from the angle of modern neuroscience as the result of self-organizing processes of meditation that can be described as attractors. In the final third part it is suggested to discern the eternity of mystical states of mind as timelessness from eternity as an attribute of the triune God.

Keywords: Time, eternity, trinity, hinduism, buddhism, islam, nunc stans, mysticism, neuroscience, attractor, chaos theory, consciousness.

1. TIME AND CONSCIOUSNESS

Ever since Plato made the fundamental distinction between time and eternity in Timaios (Ti. 37d) and Parmenides (Parm. 141aff, 151eff) the relation between time and eternity has been an issue of philosophical and religious dispute. Platon himself understood time as the movable effigy of eternity [1]. In late antiquity this debate entered into a new realm when Neo-Platonist Plotinus and Platonist St. Augustine associated time with the human consciousness. From this time onward the relation between time and eternity could be thought of as a feature of the human consciousness.

*****Address correspondence to Wolfgang Achtner:** Institute of Protestant Theology, Justus Liebig University, Giessen, Germany; Tel: +49 641 9483844; Fax: +49 641 9928009; E-mail: wolfgangachtner@gmx.de; info@wolfgangachtner.de

Plotinus (205-270)

In the Enneads III, 7 Plotinus offers a definition of eternity as independence of change and as eternal present [2]. After asking the question of what might happen, if one could partake in eternity [3], he develops a first theory of how time and eternity relate to each other within the sphere of the human soul, rejecting Aristotle's account of time [4]. According to Neo-Platonic emanation theory the soul departs from primordial unity and falls into diversity and time. It then becomes intermediate between sensuality and spirit, itself partaking also in the spirit.

There is an analogy between:

Time	Eternity
Soul	Spirit
Discursive Thinking	Experience of Unity

Time then is within the soul and its actions [5], time is the soul in movement from one phase into another [6]. However, time of the soul mirrors eternity if it tries to achieve eternity again by being on the way to wholeness, infinity, perfection and unity [7]. It can do so by concentrating on the "interior human being" and moving upward [8]. If the movement of the soul towards eternity has reached its destination by uniting with it, the time of the soul dissolves in a moment [9] of rapture in ecstasy into the eternity and timelessness of the primordial unity [10]. I would like to call this the mystic time of eternity.

St. Augustine (354-430)

Whereas Plotinus was the first to associate time intimately with the life of the soul, St. Augustine can be seen as the first one to understand time as a feature of consciousness, measured by the strength of the human soul or mind. He calls this measure of the soul's strength to maintain time "distentio animi" [11]. In his famous chapter 11 of his confessions he connects the important aspects of time such as past, present and future with consciousness and the power of the human mind that measures these aspects of time and keeps them together [12], though he does not use consciousness but animus. He thus gets a correlation in the following way (Confessions 11, 28):

Time	Consciousness/Animus
Past	Memory (praesens de praeteritis memoria)
Present	Attention (praesens de praesentibus contuitus)
Future	Expectation (praesens de futuris expectation)

However this correlation only works if the human consciousness (animus) is strong enough to transcend its location in the immediate present towards future and past in terms of memory and expectation [13]. This means that the strength of time-consciousness depends on the *extend of the mind's activity* (animus). Consciousness of time – thus interpreted – is a result of the internal activity of the soul (animus). This active part of the soul in constituting time consciousness is substantiated by the earlier writing *De immortalitate animi* (387) of St. Augustine in which he associates time consciousness with bodily activity [14]. In addition he refers to the activity of the soul, also earlier than his confessions (397-401), which is activated if one directs one's attention into the interior self as he did himself. He notes this shift of attention from the world to the self in his confessions as a necessary precondition to attain higher spheres of life [15]. He described these higher spheres of inward spiritual life – he uses the term gradatim – elsewhere before his confessions, for example in *De vera religione* (389-391) and *De genesi adversus Manichaeos* (388/390), and also in *De quantitate animae* (388) [16]. One can speculate whether or not these schemas are inspired by his early neo-platonic period of spiritual development or even by Plotinus. In any case they sound in their description rather academic and less affective, though St. Augustine mentions spiritual renascence [17], than in his highly personal confessions, where he alludes in chapter 7 to these levels of spiritual development. The interesting thing is that in both cases he relates them to the experience of time. The shift in the way time is experienced is described in *De vera religione* as a feature of the 6th level out of seven levels in which the believer is transformed into eternal life [18]. It seems however that this abstract schema of his early writings was only later enriched by his personal experience as documented in his confessions chapter 7, 17. St. Augustine describes there his ascension in the inner world from level to level (gradatim), starting with the body, the sentient soul, the superior power of the mind, the spiritual self-understanding, the changelessness (inconmutabile). Like a flash then the power of reason partakes in a moment of exaltation in pure timeless being [19]. However, St. Augustine confesses that he is not able to keep this particular moment and that he falls back again and again in temporal distractions.

He therefore confines himself to striving for this particular experience by putting his intentions to the unity of God in whom he hopes the distention of his soul will be unified [20]. It is interesting to note that this inner development of Augustine's consciousness of time corresponds strikingly with his understanding of the will. Whereas in his early neo-platonic period he adheres in his *De liberum arbitrium* (388-391) to the free will, later in chapter 8 of his confessions he relates about his tribulations concerning to cope by his will with his affections and the disunion of his will. And it is only after stabilizing his inner driving force of the will that he can analyze in chapter 11 the structure of his consciousness of time. This means that time and will are interrelated as features of the internal man.

St. Augustine and Plotinus can be understood as two different ways to experience time or to put it transcendentally to experience in time. Plotinus represents mystical way of experience time as timelessness. Augustine includes also this type of mystical experience of time, but he represents more the type of linear extension of consciousness across memory-attention-expectation. This kind of linear time within consciousness has as additional feature its directionality. There is a clear direction in the flow of time from past to present and future. However, the human spirit or consciousness, though entangled with time, also transcends time by the means of memory and expectation.

Now the interesting question is: How do these two types of time-experience relate to each other and how can they be transformed into one another? In fact we have already seen that Plotinus describes a particular moment as well as St. Augustine ("in ictu trepidantis aspectus") in which the normal way of linear time experience is transformed into a mystical one, which interestingly is closely associated with great joy.

This point of transformation from linear time to mystical time seems to be a universal property of the human mind, witnessed in all world religions. I would like to add two other examples from the Christian heritage and then proceed to the other world religions.

The first example is that of Meister Eckart in the Middle Ages. He is drawing from Augustine's neo-platonic thinking as it occurs in his *De vera religione* and he even mentions him directly in his treatise "About noble man". In particular he adopts his 6th level of time experience as entering into eternal life [21]. However he is also original in his thinking and experiencing and thus coined the notion of the Nû as the intersection of time and eternity. He is talking of the Nû (=Now) in which time dissolves into eternity [22]. It is interesting to note that Meister Eckart like St. Augustine includes the human will and its activity as a necessary precondition to attain this highest internal experience. He distinguishes two different ways of the involvement of the will in this process. First there must be an activity of the will to reach this timeless peak [23], but once it is experienced the activity of the will stops and dissolves in the eternal now [24].

Even in the rather rationalistic theology of St. Thomas Aquinas we find such mystical traces, stressing the "Now" (=nunc stans) in its role to serve as a gate to timeless eternity. "[…] quod nunc stans dicitur facere aeternitatem, secundum nostrum apprehensionem. Sicut enim causatur in nobis apprehension temporis, eo quod apprehendimus fluxum ipsius nunc, ita causatur in nobis apprehension aeternitatis, inquantum apprehendimus nunc stans" [25]. Most important also is Schleiermacher. In his talks about religion he finishes his second talk with the wonderful sentence: "In the midst of the finitude to be one with the Infinite and in every moment to be eternal is the immortality of Religion". This particular way of experience eternity within time is also described in non-religious contexts such as literature. From the English poet William Blake we do have a wonderful poem, in which this relation is expressed:

> To see a world in a grain of sand
> And heaven in a wild flower
> Hold infinity in the palm of your hand
> And eternity in an hour.

2. TIME AND MYSTICAL CONSCIOUSNESS IN THE WORLD RELIGIONS

Islam The mystical tradition in Islam had always to struggle with the idea of Allah's complete transcendence which of course entails its inaccessibility by the mystical ladder. Nevertheless in Islam a strong mystical tradition known as Sufism occurred, in which the unification with God is being sought after by a special technique of spiritual ascension, called *dikr* [26]. Islamic mysticism does know a special notion, *waqt* [27], which identifies the passage from linear time to timelessness and eternity as an attribute of Allah (*dahar*). Though it occurs in the Koran [28] in which it also has a slightly different meaning, later it becomes a special notion in the developing mystical tradition signifying the momentary transgression from linear time to the experience of timeless eternity. It is used in the manual of Islamic mysticism *Abu l-Qasim Abd al-Karim* [29]. In this eBook the notion of *waqt* [30] is used in the context of the mystical ascension of the believer to God. Therefore the Sufi master, who has attained this peak experience, is called the "son of the moment", *Ibn al-waqt* [31], free from the chains of past and future [32]. In this sense the scientist for Islamic studies, Annemarie Schimmel can characterize *waqt* as following: "The Prophet's expression 'I have a time with God' (*li ma'a Allah waqt*) is often used by the Sufis to point to their experience of *waqt*, 'time', the moment at which they break through created time and reach the Eternal Now in God [...]" [33]. This *waqt*, interpreted as 'Eternal Now', reminds very strongly to the Nû of Meister Eckharts, a comparison which is explicitly made by Annemarie Schimmel herself [34]. As an example for Islamic mysticism Ibn al-'Arabī genannt should be mentioned (1165-1240) [35], who discusses in his main work al-Futūhāt al-Makkīyah intensively the problem of time in general and *waqt* particular.

Hinduismus Within the puzzling variety of Hindu religious tradition one reasonably can expect from the mystical Yoga heritage a contribution to understanding and experiencing time and eternity after the Yoga tradition moved beyond the archaic forms of sacrifice, cults and rites. In this sense the Spanish-Indian philosopher of religion Raimundo Panikkar writes: "Later on, prana [36] is identified with immortality and brahman itself. The important point is that respiration corresponds to an internal time, and it is the mastering of this internal rhythm, especially in Yoga, that leads to the transcending of time – both externally and internally. The transition from the cultic time of the Vedas to the interiorized time of the Upanishads occurs evidently at the point where respiration, interpreted as sacrifice, takes the place of the sacrifice of fire (agnihotra). […]. The purpose of this and similar practices is patiently to succeed in discovering the unreality of time, and eventually to transcend time" [37]. It is impossible to describe the complicated entanglement of breath rhythm, nerve activity and consciousness in the various forms of Yoga. Decisive is however that Yoga, denies the independence of time, and interprets it as a feature of the activity of consciousness [38]. The philosophy of Shankhya teaches a succession of discrete atomistic units of time-consciousness which are called *ksanas*. These *ksanas* are objective in that sense, being rooted in being, as they reflect the motion of Prakriti [39]. Consciousness is attentive to these *ksanas* and constructs its succession. Thus it is emerging the construction and synthesis of the continuity of consciousness (*krama*). The continuity of time in consciousness is according to Shankhya an illusion. One can say that Shankhya and Yoga postulate a discrete theory of time. In this theory the atoms of time are discrete (*ksana*) objektiv and real (*vastu-patita*), the time-continuum however is the product of subjective construction (*krama*), it is unreal (*vastu-sunya*), and thus a deception. A specific feature of this point of view is, that in deep meditation a time-atom which is perceived with highest consciousness can open up to a kind of timeless view of all being, in which no traces of the illusion of past, present and future are being left over. This is expressed in the Patanjali-Sutra [40], a compilation of different threads of Yoga-philosophy including a commentary. The verses III, 52-54 of the Patanjali indicate the transmutation and opening of the *ksanas* to experiencing of redeemed all-temporality. "A recognition, which is redeemed, which has everything as an object and all aspects of objects, and does no longer know a sequence in time, is a recognition which is born by distinction" [41].

Buddhism The Buddha claims that the solution of the problem of time is not so much in theorizing but in existential experience. In this sense he demands from his followers, to realize the existential resolution of the time problem. Not abstract theoretical discussion, but existential experience is the right way to resolve it. In this sense he compares the one who asks about the nature of time with somebody who is hurt by a poisoned arrow. Such a person is eager to get rid of the poisoned arrow and does not ask where it comes from [42]. This hint of the Buddha underlines an evaluation of Buddhism in contemporary currents of interpretation. "The whole point of Buddhism may be summed up as living in the present", says

Dhiravamsa [43]. It is impossible to describe the time conception within the three major braches of Buddhism, Hinayana, Mahayana [44] and Vajrayana. However a glace on the early Buddhist understanding of being is possible which is underlying the zen-buddhism of Soto Zen. It is described in the main work Shōbōgenzō of its founder Dōgen (1200-1253) from Japan. In Buddhism becoming rather than being is understood as the basic category of reality. The basic constituent of reality is only the moment, instantaneity. Everything that is, is in its character momentarily (*yat sat tat ksanikam*). Real is only the instantaneous present in which time and being are intertwined and in which becoming and dissolving are interconnected and follow each other. This is the early Buddhist teaching of universal instantaneity of all being (*ksanikavada*) [45]. These three aspects of mutual entanglement of all being, the connection of being and time and the emphasis of the present as the ultimate reality also play an important role in the Zen-Buddhist philosophy of the Zen master Dōgen, which is a combination of Mahayana-Buddhism und Taoism

In his religious main work, the Shōbōgenzō [46] he articulates the strong relation between being and time. One can speak about time only in connection with being and *vice versa*. This connection is called "U-ji" [47], which can be translated as "being-time" [48]. It is also apparent in the human consciousness in its aspects of past, present and future [49].

However the present is privileged, because it can work as a kind of opening door for a comprehensive experience of time and being [50]. Time in its aspect of pure present is distinguished and can open up to trans-temporality. This is called *Nikon* [51].

By means of the hard training of consciousness in Zazen, which focuses on surmounting egocentrism, one can achieve trans-temporality [52]. The particularity of Zen-Buddhist timelessness or trans-temporality, which is based on the ontology of "U-ji", and the interpenetration of all being, is that in experiencing *Nikon* irreversibility of time is extinguished and a symmetry of time with regard to past and future in the enlightened consciousness occurs. This means that in Nikon there is a mutual interweaving of past, present and future [53].

We have seen that in all mystic traditions of the world religions there exists a connection between linear time and eternity and a specific moment in which by spiritual practice, which includes will power, one can enter in the realm of the latter. In part three it is offered an idea of how one could possibly interpret this transgression on the basis of a universal anthropology by means of neurophysiology described by chaos-theory.

3. TIME, ETERNITY AND NEUROPHYSIOLOGY

In the mystical traditions of all world religions it is of paramount importance to engage in spiritual practice in order to achieve the highest level of spiritual life. However different these spiritual practices are, they do have at least three features in common. *Firstly* they require a certain withdrawal of worldly entanglement by focusing on the interior. *Secondly* they require some technique of bodily exercise to strengthen the will, which is *thirdly* associated in most cases with some kind of regulation of breath. And finally this regulation of breath is related also to focusing attention.

If these exercises are undertaken on a regular basis the interior system of consciousness gradually changes as a result of a process of self-organization. We have already seen, that in St. Augustine's writings the consciousness of time is correlated to interior activity, one can add: the self-organizing activity of the soul (animus), bringing about a change in the soul from distraction to unification (attractor). As empirical studies have shown with plenty of evidence [54], experiencing time changes in many ways during meditation [55]. How can this change – especially if the experience of timelessness is made – be interpreted in terms of neurophysiology? I will try a tentative suggestion.

Presupposing that consciousness and attention, which resides in the dorsolateral prefrontal cortex (DLPC), are emergent properties of a complex self-referential and self-organizing system, such as the human brain, connected with the nervous system, the DLPC and the respiratory system, then one can argue that such a self-referential complex system can be described by chaos theory [56].

If this application of chaos theory is appropriate then it must be possible to find physiological correlates to basic features of chaos theory such as iteration, self-reference and (strange) attractors. The question is if the following relations make sense:

Features of Chaos Theory	Possible Correlates of the Interior Mystic Ascension to Chaos Theory
Iteration	Techniques of Meditation, such as enforced respiration, bodily action (yoga, running, dancing, fasting, sensory deprivation, sleep reduction)
Self-Reference; Self-Organization	Meditation/Cutting of from worldly entanglement
(strange) Attractors	Consciousness of linear time as a rather strong attractor Consciousness of timelessness as another attractor

If one can interpret the rather stable status of linear time consciousness in terms of chaos-theory as a rather strong attractor of the complex system of brain-nervous system-respiratory system (only distorted by sleep or mental disorders like depression, schizophrenia, mania) then one can argue that the transgression form the experience of linear time in everyday life to timelessness in mystical experience is describable in terms of the change of this system from one stable status maintained by an attractor to another stable status maintained by another attractor. The whole array of spiritual techniques then would have the function to destabilize the attractor of everyday life linear time to transform the system in a new stable status of timelessness. Understanding linear time as a stable status of a system maintained by an attractor entails a different understanding of memory and learning as a feature of linear time. Memory then must be understood not as representation rather than the result of how the brain has changed its inner dynamic in terms of self-organized learning processes including the establishment of new neuronal patterns [57]. As a consequence expectation is based on memory. Both are most likely localized in the neo-cortex as experiments show [58].

If this application of chaos theory to the experience of time and in time makes sense, then it should be possible to identify more precisely the parameters that are responsible for such a dramatic change of this system. In addition one may argue that mental distortions mentioned above could also be interpreted as pathological aberrations of "normal attractors" of the brain-nervous system-respiratory system unit. Some work in following this research strategy concerning the application of chaos theory and its attractors [59] to the brain [60], the nervous system [61] and pathologies [62] has already been done.

In this context especially the pioneering work of Walter J. Freeman has to be mentioned. He reformulates classical concept of "reinforced learning" based on "operant conditioning" by chaos theory and its attractors.

Freeman writes: "We use nonlinear mapping and multidimensional scaling into 2-space to identify itinerant chaotic trajectories through sequences of non-convergent attractor ruins in the attractor landscapes of brain state space. The attractors are created and modified by reinforcement learning based on classical and operant conditioning" [63].

This basic strategy, I guess, may also be applicable to time-consciousness as realized in every-day experience and in mediation. One only has to substitute "learning" by "meditation" based on some kind of itinerate stimulation of the brain, like breathing or memorizing a mantra. In particular Freeman has found that the transitions of a stable equilibrium to another one can be characterized by the transition from one attractor to another. "A major discovery is evidence that cortical self-organized criticality creates a pseudo-equilibrium in brain dynamics, that lets us model cortical mesoscopic state transitions as analogous to phase transitions in near-equilibrium nonliving systems like boiling or condensing water" [64]. Freeman has begun also to apply his new concept on the experience of time: "This new knowledge provides us with the neural correlates of consciousness and various states of awareness and sleep. Applications in neurophilosophy include reformulations of classic concepts of intentionality, causality, emotion, the perception of time, and the neurobiology of meaning, which we characterize as the ontological interrelation of an intentional system with its environment including other intentional systems" [65].

The problem is to make this application of chaos theory and attractors to time-perception more precise. Freeman mentions already the central role of the Neo-cortex in maintaining stability of the brain [66]. This leads to the question whether or not specific areas of the brain – not neglecting the interactions with other areas – can be identified to generate specific forms of the experience of time. Hans Förstl mentions three areas of the brain, which are responsible for generating the awareness of past, present and future, illustrated by their dysfunction in case of illness.

For our purpose of mystical experience of timelessness these localization of time with specific brain areas is of importance insofar as it helps to make the question of how to apply chaos-theory to the experience of time more specific and testable [67]. According to Förstl the brainstem and limbic system and the basal nucleus are in particular areas which generate the consciousness of the present by producing the neurotransmitter of acetyl-choline. Respiratory control in combination with pain sensitivity control, alertness, and consciousness in general is also located in this area. In addition as mentioned above the DLPC plays an important role for generating attention. All these functions are essential for meditating. Stimulation of this area of the brain by iterative breath activity in meditation in connection with the attention of the DLPC thus brings about more awareness of the present. So it can be argued that it is precisely the interaction between the DLPC and the center for the autonomous breath activity which changes both attention and breath in a self-referential loop. This could be empirically tested by checking if the production of acetyl-choline changes during meditation. A decrease of oxygen consumption has already been verified in numerous empirical studies in the case of meditation as relaxation [68], whereas in deeper advanced meditation an increase of oxygen consumption and physiological activity occurs [69]. In addition it becomes clear, that in case of the focusing on particular brain areas in meditation, such as the brainstem and the limbic system, other parts of the brain are neglected, like the frontal lobe, which in turn corresponds neatly with the gradual disappearance of the future in meditation. However it remains to be understood, how this initial attention and iterative stimulation of this part of the brain leads finally to the experience of eternity as timelessness, cut off from the attention to past and future. Since we have seen that the activation systems of alertness and consciousness, breathing and awareness are all located in the brainstem one could argue that a distortion of the fine-tuned mutual interaction of inhaling and exhaling neurons in the brainstem by meditation could lead to an integration into a more wide spread control system with more complex iterative processes of the brain [70]. In any case the iterative process has also its phenomenological equivalence in a gradual increase of awareness and concentration connected with specific ways to experience time [71], such as in quiet sitting (shrinking of time), meditation (stretching of time) absorption (now), and finally kensho (eternity), the latter leading to the experience of timelessness [72]. In this level oriented model we encounter an increase of intensity of consciousness which fits well to an increase of internal activity as exhibited in an increase of energy (oxygen) consumption [73]. This point of view is also underlined by experiments that show a global gamma coherence (see contribution of U. Ott in this volume). If one accepts that the phase transition from linear time to eternity as timelessness corresponds with the embedding of the respiratory system into a higher control system of the brain, when it resonances with it, associated with a higher degree of consciousness, then one could interpret the experience of a timeless now as the phenomenological equivalent of a resonance catastrophe.

4. TRINITY

So far we have discussed eternity understood as timelessness from an anthropological angle, arguing that it must be understood as a universal property of the self-organizing brain, which brings about the mind as an emergent trait being connected with different modes of time including eternity as a high level state.

From this point of view it should have become clear, that eternity from a theological point of view can not be identified with mystical timelessness, though God's eternity as the creator certainly also is opposed to mundane time and in this sense is timeless. Taking serious that to God the creator eternity belongs as an essential feature, one must come to the conclusion, that there must be an essential relation with the Trinitarian concept of God, his eternity, and as a result of his acting, with time. In addition, if it is true that the Jewish-Christian God has a relation to his creation and chosen people then there must be a connection between divine creativity and mundane time, not to any time but to that time, which is formed in a particular sense by God's activity, which is the history of Jews and Christians (Heilsgeschichte). Thus a necessary ingredient of such a relation between election, history and the category of novelty is Trinitarian thinking.

This connection between eternity and trinity lacks sophisticated theological elaboration in traditional theology. In fact, the basic concepts of eternity in

Western theology are not associated with Trinitarian thinking. For instance the classical definition of eternity stemming from Boethius (475-525) defines eternity as opposed to the deficient mode of human experience of time. "Eternity is the entire and complete possession of unlimited life, which becomes clear in comparison with temporality" [74], Consolatio V, 6. It is not a coincidence that time is defined by Boethius in this way, because this definition is not the result of his reasoning about time and eternity but serves the function of making intelligible that divine foreknowledge and human free will are compatible. The argument is, that the divine spirit as the entire and complete possession of unlimited live has an instantaneous knowledge of all mundane actions, without being limited by time in terms of past, present, and future, as human knowledge is. Therefore there is no divine foreknowledge at all, because he encompasses all knowledge instantaneously. For this reason the free human will and divine knowledge are compatible. However this historical context of the free-will debate in antiquity was overlooked in the theological tradition and its definition of eternity was disassociated when it was incorporated by the theological heritage. This is for example the case in St. Thomas Aquinas. Drawing from both Boethius and Aristotle, St. Thomas Aquinas [75] (1225-1274) also defines eternity as a kind of divine alternative to the deficiencies of time in the mundane world.

"Eternity is in its essence a consequence of immutability like time in its essence is a consequence of motion, as it has become clear from the aforementioned. Because God is most of all immutable, therefore he is most of all eternal" [76].

However, if the Platonic understanding of eternity as timelessness and as a divine mode of overcoming mundane deficiency of time is true, there is no way to think about an essential relation between eternity and time. But if eternity has to be associated with the creative triune God as creator, sustainer, redeemer and the one who consummates time, then this means that firstly eternity has to be redefined and secondly that there must be relations between the different triune persons and their relations to different modes of mundane time.

This understanding of time from a Trinitarian perspective was first partly elaborated by Karl Barth in his Church Dogmatics [77], after he had analyzed very astutely the relation between time and eternity [78]. However his account was only addressing the father and the son in their relation to time. He omitted completely the Holy Spirit. This theological way of thinking was again taken up by Ingo Dalferth [79]. He included the Holy Spirit and related it to the diversity of times, as are apparent in different physical theories, whereas Michael Welker started with the Holy Spirit attributing to him certain biblical traits like history or the liberation from bondage [80]. Antje Jackelen also tried to associate Trinitarian thinking with time, but finally rejected that Trinitarian thinking could be meaningfully related to the time-eternity problem [81].

However one gets the impression of all these attempts that they deal with some kind of speculative Trinitarian mathematics.

If one starts from biblical witness and the traditional dogmatic method of the "ordo cognoscendi" one must look upon the works of the Spirit, as the guiding power of God operating in the world. Doing this reveals very quickly that the Spirit as one part of the Trinitarian God is related to its operation in biblical history and its account of the history of salvation (Heilsgeschichte) and the history of contingent new events. Thus one can say that the Spirit is the driving force of God in history, which brings about change and novelty. The son of the redeemer is the power of God, which works against the decay in the world, finally in overcoming death in his crucifixion, whereas God the creator is related to the creation of the timely character of the world and its different time-structures.

Persons of Trinity	Time Structure
Holy Spirit	Driving force of biblical history (Heilsgeschichte) Creative Power to create something new in the historical process
Christ	Redeeming Time Renewing Time
Father	Creating time Maintaining Time

Eternity thus understood is not timelessness in a mystical sense or the compensation of the mundane deficiencies of time as in traditional concepts of eternity in theology, but one has to understand eternity as the simultaneous intersections of divine operations creating, maintaining, redeeming and renewing and pushing time forward and creating new times in history. This is a rather complex configuration of divine actions due to the traditional theological insight that God operates in all his persons simultaneously (opera ad extra sunt indivisa). This concept is very tentative and needs to be elaborated in more detail.

One aspect of understanding biblical history as guided by the Spirit, in particular by the work of the prophets as being the chosen actors of the divine election, is the occurrence of contingent novelties which become cornerstones to substantiate subsequent historical processes (for example the new understanding of creation in second Jesaja). There is only a tiny track of divinely guided history within general history. Therefore there is no way of prediction of these novelties, because they are not consequences of contemporary currents. Although there is no intellectual conceptualization of the work of the Spirit, generating novelty – Pannenberg's [82] claim to identify the Spirit with a field is absolutely misguided and is a setback compared with the insights of the Reformation to combine the Spirit with the preached word as opposed to the Stoic naturalism of the Spirit – one can argue that some kind of involvement for human actors in these historical processes is possible by sharing the perspective of the Spirit [83]. Partaking in this perspective means not to understand future from the perspective from the present but the present from the perspective of the future. Such a perspective overcomes the theoretical constrains of physical theories like classical mechanics and even chaos theory that still operate from a focus on the present conditions. It is only from a higher level of the Spirit that one can understand the present from the future and real novelties occur. Thus understood divine Spirit and the novelties generated in time by it never can be conceptualized in terms of an immanent development like in modern theories of emergence. Thus the idea of a perspective can be used as a link to compare in a differentiated and meaningful way the outlook on history, future and novelty both in science and Christian theology. In the following schema it turns out that there is a continuous broadening of these perspectives. Starting with the rather narrow perspective of simple classical deterministic outlook to the one can see that there is a continuous decrease of control over future and a continuous increase of possible novelties. This strikingly corresponds with the traditional theological understanding that the Spirit is free (Joh. 3, 8) and leads to novel unpredictable events (Joh. 16, 13). However the contingent operation of the spirit does not necessary preclude any human involvement, because in the gift of believing and hoping and thus sharing a particular Christian perspective the directionality of causality from future to present as working in the Spirit is foreshadowed. Thus hoping for something new, which is not yet realized, can change in a causal manner the present. Therefore in a certain sense, believing can operate as a kind of turn around of causal directionality. This of course is not possible in a setting of naturalism or nature but only in the realm of real contingencies which is human history, whereas in natural history and the evolutionary process this kind of causality is not possible, because in this realm the operating forces are that of statistical and self-referential structure.

Structure of Perspective	Structure of Future	Novelty	Computability
Simple Systems: Classical mechanics	Deterministic Present determines future	None	Absolute Computability (exception: Three body problem)
Correlated Systems: Quantum mechanics	Statistical determinism Present determines future	None	Probability of events
Recursive Systems: Chaos Theory	Deterministic, but not predictable Present determines future	Adaptive novelty	Though deterministic, not predictable
Spirit	Not deterministic, not predictable Future determines present	Absolute Novelty	No computability in eschatology. Computability in apocalyptic thinking is misguided.

Thus the future plays a decisive role in the Judeo Christian tradition, understood as governed by the Spirit, an inaccessible transcendent force, rooted in the Trinitarian God, but being foreshadowed in believing and hoping, sometimes in such a intense way that an elected human being like a prophet is dignified to be included in divine history.

The Christian understanding of time and eternity from a Trinitarian perspective shows some remarkable peculiarities in comparison to other world religions. It seems that most world religions have some particular focus on their understanding and religious esteem of time. Whereas natural religions have a focus on the past by ancestor worship, eastern religions with their mystic approach have a focus on the present favoring to delve into mystical timelessness, Judaism and Christianity have a focus on future, in which present and past however are included.

ACKNOWLEDGEMENT

Declared none.

CONFLICT OF INTEREST

The author(s) confirm that this chapter content has no conflict of interest.

REFERENCES

[1]   Timaios 37 d.
[2]   Plotinus defines eternity in terms of its independence of change. "Like how in a point everything is gathered and does not occur in flow, so remains eternity in itself and does not change, but is always in present, because nothing in it is bygone and nothing in it will be, it is only what it is", Plotinus, Enneads III, 7, 3, 18-22. "Whatever neither was nor will be, but only is what this being is as maintaining, because it does not change in what will be nor has changed, this is eternity. Thus it follows as eternity what we are seeking: Life in being which occurs in being, which is simultaneously whole, fulfilled and completely non-extended", Plotinus Enneads III, 7, 3, 33-38. "If one wants to say eternity is perfected-infinite life by means of its wholeness and does not waste anything of itself, because nothing of it is bygone or will be – otherwise it would not be whole – then one would be near to a definition", Plotinus, Enneads III, 7, 5, 25-28.
[3]   Plotinus, Enneads, III, 7, 5, 8-13. "We also must partake in eternity. But how is this possible as we are in time ", Plotinus, Enneads III, 7, 7, 5-6.
[4]   Plotinus, Enneads, III, 7, 8-9.
[5]   "First of all the soul gets into time and begot time and owns it simultaneously by its own actions", Plotinus, Enneads, III, 7, 13, 45-47.
[6]   Plotinus, Enneads III, 7, 11, 44.
[7]   Plotinus, Enneads, III, 7, 11, 43-63.
[8]   Plotinus, Enneads, VI, 9, 7, 17.
[9]   Plotinus, Enneads V 3, 17, 19; V 5, 7, 34; VI 8, 18, 8.
[10]   Plotinus, Enneads, III, 7, 12, 19-25.
[11]   "Inde mihi visum est nihil esse aliud tempus quam distentionem: sed cuius rei, nescio, et mirum, si non ipsius animi", St.Augustin, Confessions, XI, 26. Dewart, J. Mc W., 1986, 1987, 467-482; Müller, C., 1993, 123-127.
[12]   Referring to conf. XI, 26, St. Augustine claims: "In te animus meus, tempora metior", conf. XI, 27.
[13]   Sed quomodo minuitur aut consumitur futurum, quod nondum 'est', aut quomodo crescit praeteritum, quod iam non 'est', nisi quia in animo, qui illud agit, tria sunt? Nam et expectat et adtendit et meminit, ut id quod expectat per it quod adtendit transeat in id quod

meminerit. Quis agitur negat future nondum 'esse'? Sed tamen iam est in animo expectation futurorum. Et quis negat praeterita iam non 'esse'? Sed tame nest adhuc in animo memoria praeteritorum. Et quis negat praesens tempus carere spatio, quia in puncto praeterit? Sed tamen perdurat attention, per quam pergat abesse quod aderit". St. Augustin, conf. XI, 28.

[14]   Müller, C., 1993, 124f., St. Augustine, De immortalitate animae, "Porro quod sic agitur, et expectatione opus est, ut peragi, et memoria, ut comprehendi queat, quantum potest. Et expectation futurarum rerum est, praeteritarum vero memoria. At intentio ad agendum praesentis est temporis, per quod futurum in praeteritum transit, nec coepti motus corporis exspectari finis sine ulla memoria. Quomodo enim expectatur, ut desinat, quod aut coepisse excidit aut omnino motum esse? Rursus intention peragendi, quae praesens est, sine expectatione finis, qui futurus est, non potest esse; nec est quicquam, quod aut nondum est aut iam non est. Potest igitur in agendo quiddam esse, quod ad ea, quae non sunt, pertineat" (CSEL 89, S. 104f).

[15]   "Admonished to retreat to myself, I entered guided by you, into my interior self" ("Et inde admonitus redire ad memet ipsum intravi in intima mea duce te et potui", Confessions VII, 10.

[16]   Chapter 33, Quantum valeat anima?.

[17]   "sed renascuntur interius", St. Augustin, De vera religione, XXVI, 49, 133.

[18]   "The 6th brings about complete transformation into eternal life (mutationis in aeternam vitam). Now he arrives at the entire forgetting of temporal life (ad totam oblivionem vitae temporalis) [...]", St. Augustin, De vera religione, XXVI, 49, 135.

[19]   "[...] et pervenit ad id, quod est in ictu trepidantis aspectus", St. Augustin, confessions VII, 17. This delightfull moment of timelessness may be similar to Plotinus' famous rapture.

[20]   St. Augustin, conf. XI, 29.

[21]   Here is a comparison: St. Augustin in *De vera religione XXVI, 49, 135*: "sextem omnimodae mutationis in aeternam vitam et usque ad totem oblivionem vitae temproalis transeuntem, perfecta forma quae facta est ad imaginem et similitudinem dei", Meister Eckhart: "sô der mensche ist entbildet und übcrbildet von gôtes êwicheit und komen ist in ganze volkomen vergezzenlicheit zerganclîches und zîtliches lebens und gezogen ist und übergewandelt in ein götlich bilde, gotes kint worden ist", cf. Ruh, K., Vol. I, 1990, p. 94.

[22]   In his treatise "Vom edlen Menschen (About noble man)", Meister Eckart develops a system of mystical ascension, drawing from St. Augustin, identifying 6 levels. At the top, the 6th level, one can experience the "eternity (êwicheit)". In various sermons he talks about the merging of temporal man with eternity in this particular moment of the Nû (= now), like in sermon Nr. 2, 5B, 15, 38, 50, 69 Quint, J., [5]1978. For example sermon Nr. 2: "God is in this energy like in an *eternal now*. Would the mind always be united with God in this energy, man could not grow old. Because the eternal now in which God created the first man and the eternal now in which the last man will disappear and the eternal now, in which I talk, these are the same in God and only one eternal now. See, this man lives in one light with God. Therefore in him there is no suffering and sequence of time, but a constant eternity ", Quint, J., [5]1978, p. 162.

[23]   "Whenever this will turns from itself and all creatures away in a single moment into its origin, then it becomes right and free again. And in this moment all lost time is recovered", sermon Nr. 6, Quint, J., 1978 p. 181.

[24] "Wherever God is to be born in the soul, all time needs to be removed or lost with its will and desire", sermon Nr. 38, Largier, N., 1993, 409.

[25] St. Thomas Aquinas, STh. I qu. 10, art. 2.

[26] The Sufi al-Qushayrī distinguishes four groups of human beings. Those who are oriented towards the past (*ashāb as-sawābiq*), those who are oriented towards the future (*ashāb al-'awāqib*), those who are oriented towards the present (*waqt*) and finally those who are determined by the truth of God (*dhikr al-Haqq*).

[27] Islam knows many other words for time and different aspects of time, Gerhard Böwering, "The Concept of Time in Islam" in Proceedings of the American Philosophical Society, Vol. 141, No. 1. 1997, S. 58-59. Interesting to note, that the most common word for time, *zaman*, is not used in the Koran, also the word for eternity, *qidam* is does not occur.

[28] Sure 15, 38; 38, 80.

[29] "Waqt may have no reality within temporality, but it is also the ultimate reality, which, to ordinary people, would appear to be in time but which is truly timeless" (Franz Rosenthal, 1995, 23).

[30] Cf. the discussion about *waqt*: F. Meier, Abū Sa'īd-i Abū l-Hayr, Texts et Mémoires, Leiden, Teheran, Liège, S. 105-109.

[31] "The Sufi is the ›Son of the now‹, [.] the Sufi is delved in the light of the divine majesty, not the ›son‹ of anything, but free from times and states".Annemarie Schimmel, Mystische Dimensionen des Islam. Die Geschichte des Sufitums, p. 190.

[32] "Breaking through to eternity, the mystics relive their waqt, their primeval moment with God, here and now, in the instant of ecstasy, even as they anticipate their ultimate destiny. Sufi meditation captures time by drawing eternity from its edges in pre- and post-existence into the moment of mystical experience", G. Böwering, op. cit. p. 61.

[33] Annemarie Schimmel, Mystical Dimensions of Islam, The University of North Carolina Press, Chapel Hill 1975, p. 220.

[34] "Es ist das Wort *waqt*, wörtlich ›Zeit‹, das dann den ›gegenwärtigen Moment‹, den Augenblick, da dem Sufi ein gewisser Zustand geschenkt wird, ja, geradezu den *kairos* bezeichnen kann – oder in mittelalterlicher deutscher Terminologie das ›Nu‹. »Zeit ist ein schneidendes Schwert« ". Annemarie Schimmel, op. cit. p. 190.

[35] Vgl. G. Böwering, "Ibn al-'Arabī's Concept of Time", in: Gott ist schön und er liebt die Schönheit (Festschrift für Annemarie Schimmel), ed. A. Giese und J.C. Bürgel, Bern 1994, S. 71-91.

[36] Prana is the universal power of life.

[37] Raimundo Panikkar, Time and History in the Tradition of India: Kala and Karma. In : H.S. Prasad, Time in Indian Philosophy, Dehli 1992, p. 29.

[38] A critical exposition of the philosophy of time in shankhya: S.K. Sen, Time in Sankhya-Yoga. In: H.S. Prasad, op. cit. p. 505-525.

[39] Prakriti can be interpreted as matter.

[40] Written by Patanjali, the founder of the philosophy of yoga, ca. 200 BC. Vgl. B. Bäumer, B., (ed.)1979.

[41] Patanjali III, 54.

[42] Meisig, K., 1995, p. 63-68.

[43] Dhiravamsa, 1977, 33.

[44] Mahayana-Buddhism is subdivided in Madhayamika, which was founded bei Nāgārjuna (ca. 200/300 B.C.) and Yogachara, represented by Maitreyanatha in university of the

monastry Nalanda. Both teach different theories of time, which can be called ontological in the case of Nāgārjuna, in the case of Maitreyanatha idealistic in the context of his philosophy of consciousness. Further information about Nāgārjuna's conception of time: Walleser, M., vol 2, 1911, p. 111ff, vol. 3, 1912, p. 124ff. More information about Maitreyanatha: Izutsu, T., 1978, p. 309-340. The ontological strand of thought of Madhyamika and the idealistic one of Yogacara are united in the Avatamsaka-Sutrawhich is the climax of the metaphysics of Mahayana. It is also the conclusion of the development of Mahayana in India and prepares the dissemination of Buddhism to China. The Avatamsaka-Sutra is of paramount importance for the origin of Zen-Buddhism.

[45]   It is obvious that such an extreme position of pure present challenges philosophical categories like causality, memory, substance, recognition *etc.* In fact the struggle about these issues has dominated the philosophy and history of religion in India for many centuries, cf., Balslev, A.N., 1999, 91ff.

[46]   Dōgen Zenji, Vol. I, [4]1995.

[47]   Dōgen Zenji, [4]1995, p. 91-94.

[48]   "»being-time« means that being is time, time is existence, existence is time", Dōgen, 1995. p. 91.

[49]   "Do not regard time only as passing; do not examine the flowing aspect of time. If time really hastened this would be a seperation between time and us. If you believe that time is only a passing appearance, you will never understand 'time-being'. The pivotal meaning of 'being-time' is: All creatures in the whole world are cognate and con not be separated from time. Being is time and therefore it is my own true time", Dōgen, [4]1995, vol. I, p. 92.

[50]   "Each moment contains the whole world. If we understand this then this is the begin of the excersise und satori", Dōgen, [4]1995, p. 91f. "The eternal present contains the infinite space, outside nothing exists", Dōgen, [4]1995, p. 93.

[51]   "We are always living at the intersection of the horizontal and vertical dimension, that is, between temporality and trans-temporality. *Nikon*, the absolute now, is nothing but the now realized at this intersection". Abe, M., 1992, p. 100.

[52]   "It occurs by cutting through the horizontal dimensions of time in terms of the concentrated meditative practice of Zazen", Abe, M., 1992, p. 100.

[53]   "In other words, with the realization of no-self at the absolute present as the pivotal point, past and future are realized in terms of their mutuality and interpenetration, that is, their reciprocity and reversibility", Abe, M., 1992, 101.

[54]   Survey on empirical studies about meditation and their theoretical interpretation, cf., Engel, K., 1999.

[55]   Bray, J.D., 1989, Chihara, T., 1977, Chihara, T., 1989, Crowe, R.L., 1989, Handmacher, B.H., 1978, MacRae, J.A., 1983, Sudsang, R., Chentanez, V., and Veluvan, K., 1991, Sussman, A.R., 1987, Taylor, E.I., 1996, Tooley, G.A., Armstrong, S.M., Norman, T.R., and A. Sali, A., 2000.

[56]   General introduction into chaos theory, Achtner, W., 1997.

[57]   Edelman, G.M., 1993, 344-351.

[58]   Eccles, J.C., 31984, 367-370; Ingvar, 1985.

[59]   Concerning the attractors: Freeman, W.J., (2005), 92: 350-359.

[60]   Horner, H., [2]1990, p. 275-282.

[61]   Concerning the self-organization of the nervous system: Freund, H. J., [2]1990, p. 201-215.

[62]   Concerning psychotic states: Heimann, H., [2]1990, p. 215-227.

[63]   http://sulcus.berkeley.edu/.

[64]   *ibid.*

[65]   *ibid.*

[66]   "Neocortex is unique among cortices in maintaining global self-organized criticality, in which the critical order parameter is the global level of neural synaptic interaction that everywhere locally is homeostatically regulated by neural thresholds and refractory periods", http://sulcus.berkeley.edu/.

[67]   Förstl, H., "Parts of the Brain Represent Parts of the Time. Lessons from Neurodegeneration", p. 3-4; 6-7.

[68]   The physiology of breathing is a rather complicated process. Selection of literature: Austin J.H. ^{2}1998, 93-99; Block, S., *et al*, 1991; Von Euler 1986; Freeman, D., *et al.* 1981, Benson, H., *et al.*, 1977a.

[69]   Benson, H., *et al.*, 1990.

[70]   Neurons controlling in a complex interplay inhaling and exhaling, cf.: Birbaumer, N., Schmidt, R.F., 62006, 204-205.

[71]   Different ways to experience time in meditation are compiled in Marshal, P., 2005.

[72]   Austin, J.H., 1998, 561-567. This interpretation with chaos theory as a tool to understand the dynamic of a system differs from the approach of James H. Austin, who holds that the disappearance of time in meditation is due to the breakdown of time as a construct of our brain based on the interweaving of different spheres and experiences in the brain. "[…] our sense of time would seem to extend through much of the whole brain, involving regions on both sides that function in an integrated manner. Timelessness is *letting go* of all this. Through a process of transient disconnections, jammings, or bypassing", Austin, J.H. ^{2}1998, 567.

[73]   There are many models that try to identify various levels of consciousness. For example: Davidson, R.J., 1975; Alexander, Ch. N. and Boyer, R.W. 1989; Alexander, Ch. N., *et al.* 1990.

[74]   "Aeternitas est interminabilis vitae tota simul et perfecta possessio, quod ex collatione temporalium clarius liquet".

[75]   STh I qu. 10, art. 1.

[76]   "Respondeo dicendum quod ratio aeternitatis consequitur immutabilitatem, sicut ratio temporis consequitur motum, ut ex dictis patet. Unde, cum Deus sit maxime immutabilis, sibi maxime competit esse aeternum. Nec solum est aeternus, sed est sua aeternitas, cum tamen nulla alia res sit sua duration, quia non est suum esse. Deus autem est suum esse uniforme, unde, sicut est sua essential, ita est sua aeternitas", S.Th. I qu. 10, art. 2.

[77]   KD II, 1, 694-722. „On the contrary, the fact that God has and is Himself time, and the extent to which this is so, is necessarily made clear to us in His essence as the triune God", CD II, 1, 615 (Original in German: KD II, 1, 694).

[78]   "To assert the reality of time in the face of an in spite of these difficulties without the desire or the ability to set them aside, or even without letting oneself be worried by them, is perhaps in practice only possible for theology when it is revelation theology, and as such in a position to reckon not only with these two times, but in addition, with a quite different time", CD I, 2, 49. (Original in German: KD I, 2, 54).

[79]   Dalferth, I., U., 1994, 9-34.

[80]   "Through the activity of the Spirit, certain constellations of creatures are again and again torn from certain constancies and historical processes of development in salvific ways and

led into new continuities and historical processes of development in corrective and healing manners. Through the Spirit, God's creative powers are mediated and become known as saving and renewing powers that, without interruption, act upon and through creatures", Welker, M., 1998, 326.

[81]   "Der Versuch, möglichst exakte trinitarische Modelle zu entwickeln, um mit ihrer Hilfe das Verhältnis von Gott, Zeit und Ewigkeit zu erklären, erweist sich demnach nicht als der richtige Weg", Jackelén, A., Zeit und Ewigkeit. Die Frage der Zeit in Kirche, Naturwissenschaft und Theologie, Neukirchen-Vluyn, 2002, 267.

[82]   Pannenberg, W., 1996, 257-260.

[83]   The application of the notion of perspective goes back to Dietrich Ritschl's application to the biblical stories as elaborated in: Ritschl, D., 1984. It is now widely discussed in theology, for example: Dalferth, I., 2004.

LITERATURE

Abe, M., A Study of Dōgen: His Philosophy and Religion. Steven Heine (ed.), New York 1992.

Achtner, W., Die Chaostheorie, Geschichte, Gestalt, Rezeption, Berlin: Evangelische Zentralstelle für Weltanschauungsfragen, EZW-Text 135, 1997.

Alexander, Ch.N., Boyer, R.W., "Seven states of consciousness": Modern Sciente and Vedic Science 2, 4: 325-371.

Alexander, Ch.N. *et al.*, "Growth of higher states of consciousness: The Vedic psychology of human development", Alexander, Ch. N. and Langer E.J. (eds.) Higher states of human development: Perspectives on adult growth (286-340), New York, Oxford University Press, 1990.

Augustinus, De immortalitate animi, (CSEL 89).

Austin, J.H., Zen and the Brain, The MIT Press, Cambridge, Massachusets, 1998.

Balslev, A.N., A Study of Time in Indian Philosophy, New Delhi, 1999.

Bäumer, B., (ed.), Patanjali. Die Wurzeln des Yoga. Die Yoga Sutren des Patanjali mit einem Kommentar von P.Y. Deshpande, Bern, München, Wien 1979.

Barth, K., Dogmatik, KD II, 1, Zürich 1958.

Barth, K., Dogmatik, KD, I, 2, 1960.

Benson, H., Dryer, T., Hartley, L., "Decreased Oxygen Consummation at Fixed Work Intensity with Simultaneous Elicitation of the Relaxation Response", Clinical Research 25 (1977a), 453A.

Benson, H. *et al.*, "Three Case Reports of the Metabolic and Electroencephalographic Changes During Advanced Buddhist Meditation Techniques", Behavioral Medicine 1: 90-95.

Berlinger, R., "Zeit und Zeitlichkeit bei A.Augustinus", in: Zeitschrift für philosophische Forschung 7 (1953), 493-510.

Birbaumer, N., Schmidt, R.F., Biologische Psychologie, Berlin, ⁶2006.

Block, S., Lemeignan, M., Aguilera, N., "Specific respiratory patterns distinguish among human basic emotions", in: International Journal of Psychophysiology 1991; 11: 141-154.

Bray, J.D., "The Relationship of Creativity, Time Experience and Mystical Experience." Dissertation Abstracts International 50, no. 8-B (1989): 3,394.

Chihara, T., "Psychological Studies on Zen Meditation and Time-Experience." In Psychological Studies on Zen, ed. Y. Akishige. Tokyo: Zen Institute of Komazawa University, 1977.

Chihara, T., "Zen Meditation and Time Experience." Psychologia 32, no. 4 (1989): 211-220.

Crowe, R.L., "Time Perception and Hassles Appraisal in Beginning Meditators and Non-Meditators." Dissertation Abstracts International 50, no. 9-B (1989): 3916.

Dalferth, I., U., 1994. „Gott und Zeit". In: Religion und Gestaltung der Zeit, D. Georgi, H.-G. Heimbrock und M. Moxter (ed.), Kampen: Kok Pharos. p. 9-34.

Dalferth, I., U., Stoellger, P., (ed.), Wahrheit in Perspektiven, in: Religion in Philosophy and Theology 14, Tübingen 2004.

Damasio, A.R., Ich fühle, also bin ich. München 2000.

Davidson, R.J., "The Physiology of Meditation and mystical states of consciousness". Perspectives in Biology and Medicine 19: 345-380.

Dewart, J. Mc W., "Augustine's Struggle with Time and History", in Congresso internazionale su S. Agostino nel XVI centenario della conversione, Rom, 1986, Atti, Bd. 2, Rom 1987, 467-482.

Dhiravamsa, The Way of Non-Attachment. The Practice of Insight Meditation. New York, Schocken, 1977.

Dōgen Zenji, Shōbōgenzō. Die Schatzkammer der Erkenntnis des Wahren Dharma, vol. I, Zürich, München, Berlin 1995.

Duchrow, U., "Der sogenannte psychologische Zeitbegriff Augustins im Verhältnis zur physikalischen und geschichtlichen Zeit": in: Zeitschrift für Theologie und Kirche 63 (1966) 267-288, 284.

Eccles, J.C., Die Evolution des Gehirns – die Erschaffung des Selbst, München, Zürich 1994.

Edelman, G.M., Tononi, G., Universe of Consciousness, New York 2000.

Edelmann, G.M., Unser Gehirn – ein dynamisches System, München Zürich 1993.

Engel, K., Meditation. Geschichte Systematik Forschung Theorie, Frankfurt, Berlin, New York 1999.

Förstl, H., "Parts of the Brain Represent Parts of the Time. Lessons from Neurodegeneration", (2007), unpublished manuscript.

Freeman, D., Shannon, R., "Nucleus retroambigualis respiratory neurons: Responses to intercostals and abdominal muscle afferents", in: Respiration Physiology 1981; 45: 357-375.

Freeman, W.J., "Consciousness, Intentionality and Causality". In: Nuñez, R. Freeman W.J. (eds.): Reclaiming cognition, Journal of Consciousness Studies, Vol. 6, N. 11-12, 1999.

Freeman, W.J., "A field-theoretic approach to understanding scale-free neocortical dynamics", in Bio. Cybern. (2005), 92: 350-359".

Freeman, W.J., Neurodynamics. An Exploration in Mesoscopic Brain Dynamics, London 2000.

Freund, H. J., "Selbstorganisation des Zentralnervensystems", in Gerok, W., (ed.), Ordnung und Chaos in der unbelebten und belebten Natur, Stuttgart, [2]1990, 201-215.

Handmacher, B.H., "Time in Meditation and Sex Differences Related to Intrapersonal and Interpersonal Orientation." Dissertation Abstracts International 39, no. 2-A (1978): 676-677.

Heimann, H., "Ordnung und Chaos bei Psychosen", in Gerok, W., (ed), Ordnung und Chaos in der unbelebten und belebten Natur, Stuttgart, [2]1990, p. 215-227.

Horner, H., "Spin-Gläser und Hirngespinste. Einfache Modelle elementarer Funktionen des Gehirns", in Gerok, W. (ed.), Ordnung und Chaos in der unbelebten und belebten Natur, Stuttgart, [2]1990, 275-282.

Ingvar, D.H., "Memory of the future: an essay on the temporal organization of conscious awareness", in: Hum. Neurobiol 4:127-136.

Izutsu, T., "The Field Structure of Time in Zen Buddhism", in: Adolf Portmann (ed.), Eranos Jahrbuch 1978, Zeit und Zeitlosigkeit, p. 309-340.

Jackelén, A., Zeit und Ewigkeit. Die Frage der Zeit in Kirche, Naturwissenschaft und Theologie, Neukirchen-Vluyn, 2002, 267.

Largier, N., Meister Eckhart Predigten. Sämtliche deutschen Predigten und Traktate sowie eine Auswahl aus den lateinischen Werken. Kommentierte zweisprachige Ausgabe, Frankfurt 1993.

MacRae, J.A., "A Comparison between Meditating Subjects and Non-meditating Subjects on Time Experience and Human Field Motion." Dissertation Abstracts International 43, no. 11-B (1983): 3537.

Marshall, P., Mystical Encounters with the Natural World, Oxford 2005, Oxford University Press.

Meisig, K., Klang der Stille, Wien 1995.

Müller, C., Geschichtsbewußtsein bei Augustinus. Ontologische, anthropologische und universalgeschichtlich/heilsgeschichtliche Elemente einer augustinischen 'Geschichtstheorie', Würzburg 1993.

Pannenberg, W., "Geist als Feld - nur eine Metapher?", in: Theologie und Philosophie 71, 1996, 257-260.

Quint, J., Meister Eckart, Deutsche predigten und Traktate, München [5]1978.

Ritschl, D., Zur Logik der Theologie. Kurze Darstellung der Zusammenhänge theologischer Grundgedanken, München 1984.

Ruh, K., Die Geschichte der abendländischen Mystik, München 1990.

Skarda C., Freeman, W.J. (1987). "How brains make chaos in order to make sense of the world", Behavioral And Brain Sciences **10**, 161–195.

Sudsang, R., Chentanez, V., and Veluvan, K.,. "Effect of Buddhist Meditation on Serum Cortisol and Total Protein Levels, Blood Pressure, Pulse Rate, Lung Volume and Reaction Time." Physiology and Behavior 50, no. 3 (1991): 543-548.

Sussman, A.R., "Death, Time and Consciousness: A Theoretical Examination of the Psychological Impact of Alterations in Temporal Perspective." Dissertation Abstracts International 49, no. 3-B (1987): 922.

Taylor, E.I., Psychological Suspended Animation: Heart Rate, Blood Pressure, Time Estimation, and Introspective Reports from an Anechoic Environment. 2nd edition.Cambridge, MA: The Essene Press, 1996.

Tooley, G.A., Armstrong, S.M., Norman, T.R., and A. Sali, A.,"Acute Increase in Night-time Plasma Melatonin Levels Following a Period of Meditation." Biological Psychology 53 (2000): 69-78.

Von Euler, C., "Brainstem mechanisms for generation and control of breathing pattern" in: Handbook of Physiology. Sec. 3, The Respiratory System, vol. 2, pt 1, eds. A. Fishman, N. Cherniak, and J. Widdicombe. Bethesda, Md. American Physiological Society, 1986, 1-67.

Walleser, M., Die buddhistische Philosophie in ihrer geschichtlichen Entwicklung. Die mittlere Lehre des Nāgārjuna, vol. 2, Heidelberg 1911, vol 3, Heidelberg 1912.

Welker, M., "God's Eternity, God's Temporality, and Trinitarian Theology", in Theology Today 1998, p. 317-329.

INDEX

A

A-theorists 147, 150

Absolute motion 30

Absolute time 5, 15, 29-31, 119, 131, 167

   existence of 30

Abstractions 87, 119-20

Acetyl-choline 98, 100, 196

Adaptation 77, 82, 87-8

   former 87

   physiological 73, 75, 77, 81

Adapted states 72, 77-8, 80-1, 84, 86-7, 89-90

   new 77, 80

Adaptive behavior 73

Adaptive events 72-3, 76-8, 80-3, 85-9, 91

   subsequent 81, 90

   temporal vector character of 73, 90

Adaptive operation modes 72, 77-81, 86, 88, 90

Adaptive processes 77, 81, 88-9

Advanced civilisations 61

Advanced signals 170-1

Age 22, 29, 35, 39-40, 42, 48, 51, 56-7, 151

   average 21-2

   finite 41

   infinite 29, 41-4

Aging 14-15, 22, 158

Aging processes 14, 22, 25, 158

Aging universe 25

Alertness 196

Algae 81-2

Algorithm 23

Alleged ineffability 106

Alzheimer 94, 97, 99

M

Q

www.ingramcontent.com/pod-product-compliance
Lightning Source LLC
Chambersburg PA
CBHW050828220326
41598CB00006B/334